CORONA
FALSE ALARM?

Facts and Figures

Karina Reiss & Sucharit Bhakdi

Chelsea Green Publishing
White River Junction, Vermont
London, UK

Translated by Monika Wiedmann, Deirdre Anderson and Sucharit Bhakdi

Author photos: Peter Pullkowski/Sucharit Bhakdi; Dagmar Blankenburg/
Karina Reiss
Cover design: Alexandra Schepelmann/Donaugrafik.at
Layout and typesetting: Goldegg Verlag GmbH, Vienna

This edition published by Chelsea Green Publishing, 2020.

Printed in the United States of America.
First printing September 2020.
10 9 8 7 6 5 4 3 2 1   20 21 22 23 24

ISBN 978-1-64502-057-8 (paperback) | ISBN 978-1-64502-058-5 (ebook)
| ISBN 978-1-64502-059-2 (audio book)

Library of Congress Control Number: 2020945206

Chelsea Green Publishing
85 North Main Street, Suite 120
White River Junction, Vermont USA

Somerset House
London, UK

www.chelseagreen.com

For our sunshine on dark days.
Jonathan Atsadjan

Acknowledgements

The authors owe a great debt of gratitude to Monika Wiedmann for the initial translation from the German and to Deirdre Anderson for critical comments and valuable suggestions. Our heartfelt thanks to both for professional editing and proofreading of the final manuscript.

Contents

Preface

The first months of the year 2020 were characterised worldwide by a single nightmare: Corona. Dreadful images took wing from China, then from Italy, followed by other countries. Projections on how many countless deaths would occur were coupled with pictures of panic buying and empty supermarket shelves. The media in everyday life was driven by Corona, morning, noon and night for weeks on end. Draconian quarantine measures were established all over the world. When you stepped outside, you found yourself in a surreal world – not a soul to be seen, but instead empty streets, empty cities, empty beaches. Civil rights were restricted as never before since the end of the Second World War. The collapse of social life and the economy were generally accepted as being inevitable. Was the country under threat of such a dreadful danger to justify these measures? Had the benefits that could possibly be gained by these measures been adequately weighed against the subsequent collateral damage that might also be expected? Is the current plan to develop a global vaccination programme realistic and scientifically sound?

Our original book was written for the public in our country and this translated version is tilted toward the

German narrative. However, global developments have advanced along similar lines, so that the basic arguments hold. We have replaced a number of local events in favour of pressing new issues regarding the question of immunity and the postulated need for development of vaccines against the virus.

The intent of this book is to provide readers with facts and background information, so that they will be able to arrive at their own conclusions. Statements in the book should be regarded as the authors' opinions that we submit for your scrutiny. Criticism and dissent are welcome. In scientific discussions, postulation of any thesis should also invite antitheses, so that finally the synthesis may resolve potential disagreement and enable us to advance in the interest of mankind. We do not expect all readers to share our points of view. But we do hope to ignite an open and much needed discussion, to the benefit of all citizens of this deeply troubled world.

How everything started

In December of 2019, a large number of respiratory illnesses were recorded in Wuhan, a city with about 10 million inhabitants. The patients were found to be infected with a novel coronavirus, which was later given the name SARS-CoV-2. The respiratory disease caused by SARS-CoV-2 was designated COVID-19. In China, the outbreak evolved into an epidemic in January 2020, rapidly spreading around the globe (1, 2, 3).

Coronaviruses: the basics

Coronaviruses co-exist with humans and animals worldwide, and continuously undergo genetic mutation so that countless variants are generated (4, 5). "Normal" coronaviruses are responsible for 10–20% of respiratory infections and generate symptoms of the common cold. Many infected individuals remain asymptomatic (6). Others experience mild symptoms such as unproductive cough, whilst some additionally develop fever and joint pains. Severe illness occurs mainly in the elderly and can take a fatal course, particularly in patients with pre-existing illnesses, especially of heart and lung. Thus, even "harmless" coronaviruses can be associated with case fatality rates of 8% when they gain entry to nursing homes (7). Still, due to their marginal clinical significance, costly measures for diagnosing coronavirus infections are seldom undertaken, searches for antiviral agents have not been prioritised, and vaccine development has not been subject to serious discussion.

Only two members of the coronavirus family reached world headlines in the past.

SARS virus (official name: SARS-CoV) entered the stage in 2003. This variant caused severe respiratory illness with a high fatality rate of approximately 10%.

Fortunately, the virus turned out not to be highly contagious, and its spread could be contained by conventional isolation measures. Only 774 deaths were registered worldwide (8, 9). Despite this manageable danger, fear of SARS led to a worldwide economic loss of 40 billion US dollars (8). Coronaviruses subsequently faded into the background. A new variant, MERS-CoV, emerged in the Middle East in 2012 and caused life-threatening disease with an even higher fatality rate of more than 30%. But contagiousness of the virus was also low and the epidemic was rapidly brought under control (10).

China: the dread threat emerges

When the news came from China that a new coronavirus family member had appeared on stage, the most pressing question was: would it be harmless like its "normal" relatives or would it be SARS-like and highly dangerous? Or worse still: highly dangerous *and* highly contagious?

First reports and disturbing scenes from China caused the worst to be feared. The virus spread rapidly and with apparent deadly efficacy. China resorted to drastic measures. Wuhan and five other cities were encircled by the army and completely isolated from the outside world.

At the end of the epidemic, official statistics reported about 83,000 infected people and fewer than 5,000 fatalities (11), an infinitesimally small number in a country with 1.4 billion inhabitants. Either the lockdown worked or the new virus was not so dangerous after all. Whatever the case, China became the shining example on how we could overcome SARS-CoV-2.

More disturbing news then came from northern Italy. Striking swiftly, the virus left countless dead in its wake. Media coverage likened the situation to "warlike conditions" (12). What was not reported was that in other parts of Italy, and also in most other countries, the "fatality rate" of COVID-19 was considerably lower (13, 14).

Could it be that the intrinsic deadliness of one and the same virus varied, depending on the country and region it invaded? Not very likely, it seemed.

How dangerous is the new "killer" virus?

Compared to conventional coronaviruses

Gauging the true threat that the virus posed was initially impossible. Right from the beginning, the media and politicians spread a distorted and misleading picture based on fundamental flaws in data acquisition and especially on medically incorrect definitions laid down by the World Health Organization (WHO). Each positive laboratory test for the virus was to be reported as a COVID-19 case, irrespective of clinical presentation (15). This definition represented an unforgiveable breach of a first rule in infectiology: the necessity to differentiate between "infection" (invasion and multiplication of an agent in the host) and "infectious disease" (infection with ensuing illness). COVID-19 is the designation for severe illness that occurs only in about 10% of infected individuals (16), but because of incorrect designation, the number of "cases" surged and the virus vaulted to the top of the list of existential threats to the world.

Another serious mistake was that every deceased person who had tested positive for the virus entered the official records as a coronavirus victim. This method of reporting violated all international medical guidelines (17). The absurdity of giving COVID-19 as the cause of death in a patient who dies of cancer needs no comment. Correlation does not imply causation. This was causal fallacy that was destined to drive the world into a catastrophe. Truth surrounding the virus remained enshrouded in a tangle of rumours, myths and beliefs.

A French study, published on March 19, brought first light into the darkness (6). Two cohorts of approximately 8,000 patients with respiratory disease were grouped according to whether they were carrying everyday coronaviruses or SARS-CoV-2. Deaths in each group were registered over two months. However, the number of fatalities did not significantly differ in the two groups and the conclusion followed that the danger of "COVID-19" was probably overestimated. In a subsequent study, the same team compared the mortality associated with diagnosis of respiratory viruses during the colder months of 2018–2019 and 2019–2020 (week 47-week 14) in southeastern France. Overall, the proportion of respiratory virus-associated deaths among hospitalised patients was not significantly higher in 2019–2020 than the year before (18). Thus, addition of SARS-CoV-2 to the spectrum of viral pathogens did not affect overall mortality in patients with respiratory disease.

Regarding the number of deaths

How can the aforementioned be reconciled with the official reports of the horrifying number of COVID-19 deaths? Two numbers must be known if the danger of a virus is to be assessed:

the number of infections and the number of deaths.

How many were infected by the new virus?

Attempts to answer this question were beset by three problems:

1. How reliable was the test for virus detection?

The virus is present in the nasopharynx for approximately two weeks, during which time it can be detected. How is this done? Viral RNA is transcribed into DNA and quantified by the so-called polymerase chain reaction (PCR). The first assay for the new coronavirus was developed under guidance of Professor Christian Drosten, Head of the Institute for Virology at the Charité Berlin. This test was used worldwide in the initial months of the outbreak (19). Tests from other laboratories followed (20).

Diagnostic PCR tests must normally undergo stringent quality assessment and be approved by regulatory agencies before use. This is important because no laboratory test can ever give 100% correct results. The quality control requirements were essentially shelved in the case of SARS-CoV-2 because of declared international urgency. Consequently, nothing was really known regarding test reliability, specificity and sensitivity. In essence, these parameters give an indication of how many false-positive or false-negative results should be expected. The test protocol from the Drosten laboratory were used worldwide, and test results played a key role in political decision-making. Yet, data interpretation was often largely a matter of belief. What did Drosten himself say on Twitter (21)?

» Sure: Towards the end of the illness the PCR is sometimes positive and sometimes negative. Here, chance plays a role. When you test a patient twice as negative and discharge him as cured, it is indeed possible that you can have positive test results again at home. But this is still far from being a re-infection.

Several physician colleagues have informed us of similar haphazard results with patients who had been tested repeatedly during their hospitalisation. Is it particularly surprising that goats and papayas tested positive for the virus in Tanzania? The criticism by the President of Tanzania regarding the unreliability of the test kits was of course immediately dismissed by the WHO (22).

But today it is perfectly clear that the test result is error-prone, as is every PCR (23, 24). How much so, and whether there are significant differences among the presently available tests, cannot be determined because of lack of data.

So let us assume that the PCR test is incredibly good and produces 99.5% correct results. That sounds, and would indeed be, exceptional – it means that one can expect only 0.5% false-positives. Now take the cruise ship "Mein Schiff 3". After a crew member had tested positive for the virus, almost 2,900 people from 73 countries were forced into "ship quarantine". Many had been on board for nine months. Complaints reached the outside world about the "prison-like" conditions, psychological problems abounded and nerves were frayed (25).

Nine positive cases were reported after testing was completed. One person who tested positive had a cough, the other eight were without symptoms. Might they have belonged to the 0.5% false-positive cases, as perhaps the very first case had been? Where were the true-positives that must theoretically have been there? Were they possibly tested as false-negatives or were all positive tests false?

In the context of false results, we should consider the following: when the epidemic subsided (in Germany, in mid-April,) PCR testing became a dangerous source of misinformation because numbers of new cases were derived from the "background noise" of false-positive results. When all 7,500 employees of the Charité Berlin (one of Europe's largest university hospitals) were tested from April 7 to April 21, 0.33% were positive (26). True or false?

When positive test rates drop below a certain limit, it is senseless to continue mass screening for the virus in non-symptomatic individuals. And use of numbers acquired under these circumstances as a reason for implementing any measures should not be tolerated.

2. Selective or representative? Who was tested?

There is only one way to approximate how many people are infected during an epidemic with an agent that causes high numbers of unnoticed infections: at sites of an outbreak, the population must be tested as extensively as possible. But scientists who called for this during the coronavirus epidemic (27, 28) were ignored.

Instead, the Robert Koch Institute (RKI), the German federal government agency and research institute for disease control, stipulated at the beginning that only selective testing should be carried out – exactly the opposite of what should have happened. And as the epidemic ran its course, the RKI stepwise altered the testing strategy – always in the diametrically wrong direction (29).

At first, only people who had been in a high-risk area and/or had been in contact with an infected person and also presented with flu-like symptoms were to be tested. At the end of March, the RKI then changed the recommended test criteria to: flu-like symptoms and, at the same time, contact with an infected person. At the beginning of May, the President of the RKI, Professor Lothar Wieler, announced people with even "the slightest symptoms" should be tested (29).

The responsibility for translating these dubious decisions into action lay entirely within the hands of the local health authorities. A co-worker at our lab was a typical example: the coach of her handball team was coronavirus positive. The players – all from different administrative districts – were sent home on 14-

day quarantine. One player developed symptoms with coughing and hoarseness and wanted to get tested but was refused on the grounds that she had no fever. A player from a neighbouring district had no symptoms but the local health authority ordered a test despite this fact.

This resulted in chaos, caused by the appalling ineptitude of the authorities from top to bottom. What would have been urgently needed instead were scientifically sound studies to clarify basic issues of virus dissemination. As many as possible should have been tested in outbreak areas. Antibody responses in those that had tested positively could have subsequently been assessed.

Only a single such study addressing these questions was undertaken in Germany: the Heinsberg investigation conducted by Professor Hendrik Streeck, Director of the Institute for Virology at the University of Bonn. Aware of the importance of the preliminary data, these were presented at a press conference – where Streeck was torn apart by the disbelieving media (30, 31). The fatality rate was ridiculed as being impossible because it was ten times lower than what acknowledged experts and the WHO had been spreading as established facts. After completion of the study, final results essentially confirming the preliminary report were again presented, and again deemed by the media to be flawed and inconclusive. But the results of the study spoke for themselves (32) – and they contradicted the panic propaganda of the media.

3. The number of conducted tests directly influences infection statistics

A third factor added to the statistical mess. Imagine that you wanted to count the number of a migratory bird species in a large lake district. There are hundreds of thousands but your counting device can only count 5,000 per day. Next day, you ask a colleague to help, and together you arrive at 10,000 counts. The day after that, two more colleagues join in and 20,000 birds are counted. In short, the higher the testing capacity/ number of tests, the higher the numbers – as long as innumerable unidentified cases abound, as with SARS-CoV-2 (16, 32–36). The more tests are performed, the more COVID-19 cases are found during the epidemic. This is the essence of a "laboratory-created pandemic".

Now recall that the test has neither 100% specificity nor 100% sensitivity – meaning that occasionally you would mistake a log for a bird. Therefore, even after all our birds have long since moved on, you would still "find" many by just performing a sufficient number of tests.

In conclusion, no reliable data existed regarding the true numbers of infection at any stage of the epidemic in this country. At the peak of the epidemic, the official numbers must have been gross underestimates – in the order of 10 or even more. At its wane at the end of April in Germany, the numbers must also have been gross overestimates.

Basing any political decisions on official numbers at any stage was fallacy.

How many deaths did SARS-CoV-2 infections claim? Here, again, we have the dilemma of definition: what is a "coronavirus death"?

If I drive to the hospital to be tested and later have a fatal car accident – just as my positive test results are returned – I become a coronavirus death. If I am diagnosed positive for coronavirus and jump off the balcony in shock, I also become a coronavirus death. The same is true for a sudden stroke, etc. As openly declared by RKI president Wieler, every individual with a positive test result at the time of death is entered into the statistics. The first "coronavirus death" in the northernmost state of Germany, Schleswig-Holstein, occurred in a palliative ward, where a patient with terminal oesophageal cancer was seeking peace before embarking on his last journey. A swab was taken just before his demise that was returned positive – after his death (37). He might equally well have been positive for other viruses such as rhino-, adeno- or influenza virus – if they had been tested for.

This particular case did not need more testing or a post-mortem to determine the actual cause of death.

However, with the emergence of a new and possibly dangerous infectious disease, autopsies should be undertaken in cases of doubt to clarify the actual cause of death. Only one pathologist ventured to fulfil this task in Germany. Against the specific advice of the RKI, Professor Klaus Püschel, Director of the Institute of Forensic Medicine, Hamburg University, performed autopsies on all "coronavirus victims" and found that not one had been healthy (38). Most had suffered from several pre-existing conditions. One in two suffered from coronary heart disease. Other frequent ailments were hypertension, atherosclerosis, obesity, diabetes, cancer, lung and kidney disease and liver cirrhosis (39).

The same occurred elsewhere. Swiss pathologist Professor Alexander Tzankov reported that many victims had suffered from hypertension, most were overweight, two thirds had heart problems and one third had diabetes (40). The Italian Ministry of Health reported that 96% of COVID-19 hospital deaths had been patients with at least one severe underlying illness. Almost 50% had three or more pre-existing conditions (41).

Interestingly, Püschel found lung embolisms in every third patient (39). Pulmonary embolisms usually arise through detachment of blood clots in deep veins of the leg that are swept into the lungs. Clots typically form when blood flow sags in the legs, as when the elderly spend the day seated and inactive. A high frequency of lung embolisms was already described in deceased influenza patients 50 years ago (42). Thus, we are not on the verge of discovering a unique property of SARS-Cov-2 that would heighten its threat, but we do bear witness to the absurd situation where the elderly seek to protect themselves by obeying the chant that sounds around the world: "Stay at home". Physical inactivity is pre-programmed, thromboses included? Swedish epidemiologist Professor Johann Giesecke recommended exactly the opposite: As much fresh air and activity as possible. The man knows his job!

The number of genuine COVID-19 fatalities remained unknown outside Hamburg. The situation was no better in other countries. Professor Walter Riccardi, adviser to the Italian Ministry of Health, stated in a March interview with "The Telegraph" that 88% of the Italian "coronavirus deaths" had not been due to the virus (43).

The problem with coronavirus death counts is such that the numbers can be viewed as nothing other than gross overestimates (44). In Belgium, not only fatalities with a positive COVID-19 test entered the ranks but also those where COVID-19 was simply suspected (45).

Scientific competence did not seem to rule the agenda of Germany's RKI. Fortunately, there are scientists who stand out in contrast. Stanford Professor John Ioannidis is one of the eminent epidemiologists of our times. When it became clear that the epidemic in Europe was nearing its end, he showed how the officially reported numbers of "coronavirus deaths" could be used to calculate the absolute risk of dying from COVID-19 (46).

The risk for a person under 65 years in Germany was about as high as a daily drive of 24 kilometres. The risk was low even for the elderly ≥ 80 with 10 "coronavirus deaths" per 10,000 ≥ 80-year olds in Germany (column at the far right).

Calculation of this number is simple. About 8.5 million citizens are ≥ 80 years in Germany. About 8,500 "coronavirus deaths" were recorded in this age group. This leads to an absolute risk of coronavirus death of 10 per 10,000 ≥ 80 year-olds. Now realise that every year about 1,200 of 10,000 ≥ 80-year olds die in Germany (black column, data from the Federal Office of

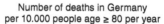

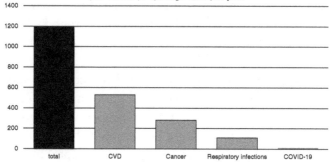

Statistics). Nearly half of them due to cardiovascular diseases (CVD), almost a third from cancer and around 10% (over 100) owing to respiratory infections. The latter have always been caused by a multitude of pathogens including the coronavirus family. It is obvious that a new member has now joined the club, and that SARS-CoV-2 cannot be assigned any special role as a "killer virus".

This is underlined by another observation. Severe respiratory infections are registered by the RKI in the context of influenza surveillance. The vertical line marks the time when documentation of SARS-CoV-2 infections was started. Was there ever any indication for an increase in the number of respiratory infections (47)? No, the 2019/20 winter peak is followed by typical seasonal decline. And note that the lockdown (red arrow) was implemented when the curve had almost reached base level.

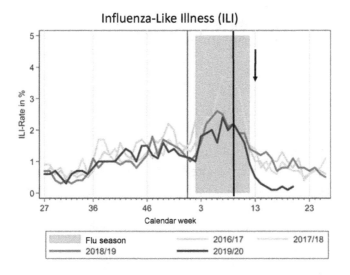

Influenza-Like Illness (ILI)

Source: Homepage RKI (Fig. 1), https://grippeweb.rki.de/

How does the new coronavirus compare with influenza viruses?

The WHO warned the world that the COVID-19 virus was much more infectious, that the illness could take a very serious course, and that no vaccine or medication was available.

The WHO abstained from explaining that truly effective medication hardly exists against any viral disease and that vaccination against seasonal flu is increasingly recognised as being ineffective or even counterproductive. Furthermore, the WHO disregarded two points that needed to first be addressed before any valid comparison of the viruses could be undertaken.

How many people die of COVID-19 compared with influenza?

The WHO claimed that 3–4% of COVID-19 patients would die, which by far exceeded the fatality rate of annual influenza (48).

This is important enough to call for a closer look. Influenza viruses pass wave-like through the population. The waves can be small in one year and high in another. Case fatality rates are 0.1% to 0.2% during a normal flu season in Germany (49), which translates to several hundreds of deaths. In contrast, there were approximately 30,000 influenza-related deaths in the 1995/1996 season (50) and approximately 15,000 deaths in 2002/2003 and 2004/2005.

The RKI estimates that the last great flu epidemic of 2017/2018 claimed 25,000 lives (51). With 330,000 reported cases, the fatality rate would be ~8% (52). As in all previous years, Germany weathered this epidemic without implementing any unusual measures.

The WHO estimates that there are 290,000–650,000 flu deaths each year (53).

Now turn to COVID-19. In May, the RKI calculated that 170,000 infections with 7,000 coronavirus deaths equals a 4% case fatality rate – as predicted by the WHO! Conclusion: COVID-19 is really ten times more dangerous than seasonal flu (54).

However, the number of infections was at least ten times higher because most mild and asymptomatic cases had not been sought and detected (55–59). This would bring us to a much more realistic fatality rate of 0.4%. Moreover, the number of "true" COVID-19 deaths was lower because many or most had died of causes other than the virus. Further correction of the number brings us to a rough estimate of 0.1% – 0.3%, which is in the range of moderate flu. This tallies well with the results of Professor Streeck, who arrived at an estimate of 0.24% – 0.26% based on the data of his Heinsberg study. The average age of the deceased who tested positive was around 81 years (32).

The conclusion that COVID-19 is comparable to seasonal flu has been reached by many investigators in other countries. In an analysis of several studies, Ioannidis showed that, contingent on local factors and statistical methodology, the median infection fatality rate was 0.27% (60). Many other investigators arrived at similar conclusions. All studies to date thus clearly show that SARS-CoV-2 is not a real "killer virus" (61–71).

Flu and COVID-19: who are the vulnerable?

Influenza viruses are dangerous mainly to individuals of ≥ 60 years but can sometimes also cause fatal infections in younger people.

A salient feature of the virus is that after its multiplication and release, it induces the infected host cell to commit suicide. This is a major predisposing factor for bacterial super-infections (72), which were the major cause of death during the Spanish flu.

In contrast, coronaviruses are inherently less destructive. Patients show characteristic changes in their lungs, but whether the virus is deadly or not depends less on the virus and more on the patient's overall state of health. Time and again, press reports appear on "completely healthy" young people who nonetheless were carried off by the virus. We do not know of a single case where it did not turn out afterwards that the person had not been "completely healthy", but rather had suffered for years from hypertension, diabetes or other illnesses that had gone undetected.

Sensational news: 103 year-old Italian woman recovers from COVID-19 (73)! In fact, she was not the only old lady who survived the infection without problems. Most actually did (74). The record is held by a 113 year-old Spanish woman (75).

Although the median age of the deceased is over 80 in Germany and other countries (41, 76–78), age *per se* is not the decisive criterion. People without severe pre-existing illness need fear the virus no more than young people. As we know from Püschel's and many other reports, SARS-CoV-2 is almost always the last straw that breaks the camel's back. While this is certainly sad for the family and loved ones, it is still no reason to assign the virus any heightened role. We need to keep in mind that every year, millions die of respiratory tract infections, with a whole spectrum of bacterial and viral agents playing causal roles.

One must not forget that the true cause of a death is

the disease or condition that triggers the lethal chain of events. If someone suffering from severe emphysema or end-stage cancer contacts fatal pneumonia, the cause of death is still emphysema or cancer (79, 80).

This basic rule is simply ignored in times of coronavirus. Even worse – once tested positive for SARS-CoV-2, (even falsely) – an individual can remain marked as a COVID-19 victim for life, depending on the inclination of the responsible authority (81, 82). Then, irrespective of when and why death occurs, he or she will enter the COVID-19 death register.

Thus, the number of coronavirus deaths will continue to soar incessantly. Fear in the general populace is further fuelled by reports that SARS-CoV-2 is much more dangerous than the flu because it attacks many different organs with probable long-term consequences. Newspaper reports and publications abound that the virus can be found in the heart, liver, and kidneys (83). It may even find its way to our central nervous system?!

Such headlines sound terrifying. However, obtaining positive RT-PCR results for SARS-CoV-2 in organs other than the lung is nothing surprising. The virus uses receptors to enter our cells that are not only on the surface of lung cells. But two issues are of decisive importance: the actual viral load and the question of whether the viruses cause any damage. The highest SARS-CoV-2 concentrations have been found in the lungs of patients – as is to be expected. Traces of the virus have been detected in other organs (83). Most probably, they bear no relevance. Until scientific evidence to the contrary is available, the findings must be left for what they are: trivial observations.

Is there a difference with the flu? No. It has been known for years that influenza can affect the heart and other organs (84, 85). All respiratory viruses can find their way to the central nervous system (86). There is no basic difference with SARS-CoV-2. Once in a while, patients may suffer from long-term consequences. This applies to all viral diseases, and they are exceptions. It is the exception that proves the rule.

What do we learn from all of this? COVID-19 is a disease that makes some people sick, proves fatal to a few, and does nothing to the rest. Like any annual flu.

Of course, it was always necessary to take special care not to bring these agents to elderly persons with pre-existing illnesses. When you feel unwell, refrain from visiting grandma and grandpa, especially if they are suffering from a heart condition or lung disease. And whoever has the flu will stay at home anyway. That is how everything has been and how everything should continue.

The fact that SARS-CoV-2 does not constitute a public danger and that the infection often runs its course without symptoms might have one disadvantage. Perhaps asymptomatic people are contagious and unknowingly pass the virus on to others. This fear originated from a publication co-authored and widely publicised by Drosten, in which it was reported that the Chinese businesswoman who infected an automotive supplier's staff member during a visit to Bavaria displayed no symptoms herself (87). This publication caused a worldwide sensation with expected effects, for a deadly virus that could be transmitted by healthy individuals was akin to a swift and invisible killer. This fear became the driving force behind many extreme preventive measures – from visiting bans for hospitalised patients all the way to obligatory mask-wearing.

In the midst of general panic, a very important fact escaped general attention. The major statement of the publication turned out to be false. A follow-up inquiry revealed that the Chinese woman had been ill during her stay in Germany and was under medication to relieve pain and reduce fever (88). This was not mentioned in the publication (87).

Another study that was published in April by the Drosten laboratory also came under international criticism. It concerned the question about the role of children in disease transmission. According to the Drosten study, asymptomatic children were just as contagious as adults. This message caused great concern to the general public and influenced subsequent decisions by the government. In fact, no studies exist to indicate that children play any significant role as vectors for transmission of this disease.

Be that as it may, there was no reason for completely pointless measures like closing schools and day care centres, which are known to do nothing to protect the high-risk groups (89). And no reason whatsoever to drive social life and the economy against the wall.

What is wrong with Germany – and this whole world?

Well, all the pictures disseminated so effectively by

the international media – from Italy, Spain, England and then even from New York – coupled with model calculations for hundreds of thousands, or maybe even millions of deaths – planted the firm conviction in the general populace: It simply HAS TO BE a killer virus!

The situation in Italy, Spain, England and the USA

Since the end of March, one sensation outdid the next: Italy had the most deaths, the fatality rate shocked us to the core; Spain surpassed Italy (in the number of infections); the United Kingdom broke the sad European record, exceeded only by the US. The press delighted in spreading as much terrifying news as humanly possible.

But let us reflect a little. The impact of an epidemic is dependent not only on the intrinsic properties and deadliness of the pathogen but also to a very significant extent on how "fertile" the soil is on which it lands. All reliable figures tell us we are not dealing with a killer virus that will sweep away mankind. So what did happen in those countries from which these dreadful pictures emerged?

Detailed answers to this question must be sought on the ground. Nevertheless, several facts are sufficiently known to warrant mention here. Problems surrounding coronavirus statistics went totally rampant in Italy and Spain. Elsewhere, testing for the virus was generally performed on people with flu-like symptoms and a certain risk of exposure to the virus. At the height of the epidemic in Italy, testing was restricted to severely ill patients upon their admission to the hospital. Illogically, testing was widely performed post-mortem on deceased patients. This resulted in falsely elevated case fatality rates combined with massive underestimates of actual infections (90).

As early as mid-March, the Italian GIMBE (Gruppo Italiano per la Medicina Basata Sulle Evidenze / Italian Evidence-Based Medicine Group) foundation stated that the *"degree of severity and lethality rate are largely overestimated, while the lethality rates in Lombardy and the Emilia-Romagna region were largely due to overwhelmed hospitals"* (91).

The fact that no distinction was made between "death by" and "death with" coronavirus rendered the situation hopeless. Almost 96% of "COVID-19 deaths" in Italian hospitals were patients with pre-existing illnesses. Three quarters suffered from hypertension, more than a third from diabetes. Every third person had a heart condition. As almost everywhere else, the average age was above 80 years. The few people under 50 who died also had severe underlying conditions (41).

The inaccurate method of reporting "coronavirus deaths" naturally spread fear and panic, rendering the general public willing to accept the irrational and excessive preventive measures installed by governments. These turned out to have a paradoxical effect. The number of regular deaths increased substantially over the number of "coronavirus deaths". The Times reported on April 15: England and Wales have experienced

35

a record number of deaths in a single week, with 6,000 more than average for this time of year. Only half of those extra numbers could perhaps be attributed to the coronavirus (92). There was a well-founded concern that the lockdown may have unintentional but serious consequences for the public's health (93).

It became increasingly clear that people avoided hospitals even when faced with life-threatening events such as heart attacks because they were afraid of catching the deadly virus. Patients with diabetes or hypertension were no longer properly treated, tumour patients not adequately tended to.

The UK has always had massive problems with its health care system, medical infrastructure and a shortage of medical personnel (94, 95). Due to Brexit, the UK also lacks urgently needed foreign specialists (96).

Many other countries have problems along the same lines. When the influenza epidemic swept over the world in the winter of 2017/2018, hospitals in the US were overwhelmed, triage tents were erected, operations were cancelled and patients were sent home. Alabama declared a state of emergency (97–99). The situation was little different in Spain, where hospitals just collapsed (100, 101), and in Italy, where intensive care units in large cities ground to a halt (102).

The Italian health care system has been downsizing for years, the number of intensive care beds is much lower than in other European countries. Furthermore, Italy has the highest number of deaths from hospital-acquired infections and antibiotic-resistant bacteria in all of Europe (103).

Also, Italian society is one of the oldest worldwide. Italy has the highest proportion of over 65 year-olds (22.8%) in the European Union (104). Add to that the fact that there is a large number of people with chronic lung and heart disease, and we have a much greater number in the "high-risk groups" as compared to other countries. In sum, many independent factors come together to create a special case for Italy (105, 106).

Since northern Italy was particularly affected, it would be interesting to ask if environmental factors had an influence on the way things developed there. Northern Italy has been dubbed the China of Europe with regard to its fine particulate pollution (107). According to a WHO estimate, this caused over 8,000 additional deaths (without a virus) in Italy's 13 biggest cities in 2006 (108). Air pollution increases the risk of viral pulmonary disease in the very young and the elderly (109). Obviously, this factor could generally play a role in accentuating the severity of pulmonary infections (110).

Suspicions have been voiced that vaccination against various pathogens such as flu, meningococci and pneumococci can worsen the course of COVID-19. Investigations into this possibility are called for because Italy indeed stands out with its officially imposed extensive vaccination programme for the entire population.

Yet despite all these facts, the only pictures that remain imprinted on our minds are the shocking scenes of long convoys of military vehicles carting away endless numbers of coffins from the northern Italian town of Bergamo.

Vice chairman of the Federal Association of German Undertakers, Ralf Michal, noted (111): in Italy, cremations are rather rare. That is why undertakers were overburdened when the government ordered cremations in the course of the coronavirus pandemic. The undertakers were not prepared for that. There were not enough crematoriums and the complete infrastructure was lacking. That is why the military had to help out. And this explains the pictures from Bergamo. Not only was there no infrastructure, there was also a shortage of undertakers because so many were in quarantine.

And finally, let us examine the United States, where only parts of the country were severely affected. In states like Wyoming, Montana or West Virginia, the number of "coronavirus deaths" was a two-digit figure (Worldometers, middle of May, 2020).

The situation in New York was different. Here, doctors were overwhelmed and did not know which patients to treat first, while in other states, hospitals were eerily empty. New York was the centre of the epidemic, where more than half of the COVID-19 deaths nationwide occurred (date: May 2020). Most of the deceased lived in the Bronx. An emergency doctor reported (112): "These people come way too late, but their reasoning is understandable. They are afraid of being discovered. Most of them are illegal immigrants without residence permits, without jobs and without any health insurance. The highest mortality rate is recorded in this group of people".

It would be of interest to learn how they were treated. Were they given high doses of chloroquine as recommended by the WHO? About a third of the Hispanic population carries a gene defect (glucose-6-phosphate dehydrogenase) that causes chloroquine intolerance with effects that can be lethal (113, 114). More than half of the population in the Bronx is Hispanic.

Countries and regions can differ so widely with respect to a myriad of factors that a true understanding of any epidemic situation cannot be obtained without critical analysis of these determinants.

Corona-situation
in Germany

The German populace should have been reassured that this country was well-positioned and that disturbing scenarios similar to those seen in northern Italy or elsewhere need NOT be feared. Instead, the exact opposite happened. The RKI issued warning after warning, and the government embarked on a crusade of fearmongering that defied description. Anyone who dared to challenge the warning that the world was facing the greatest pandemic threat of all times was defamed and censored.

The indicators for when which measures were supposedly necessary or no longer necessary changed haphazardly according to demand. At the beginning of March, it was the doubling rate for the numbers of infections which at first should exceed 10 days; but when this "goal" was reached, the rate had to be further slowed to 14 days. This objective was also quickly achieved so a new criterion had to be issued: the reproduction factor ("R"), which supposedly told us how many people became infected by one contagious person. The authorities at first decided that this number must decrease to less than 1. When this happened – in mid-March

– they ran into difficulties and set out to re-direct the number upward by increasing the numbers of tests. At the end of May, a bit of creative thinking led to the idea of defining a critical upper limit to the acceptable number of daily new infections: 35 per 100,000 citizens in any town or region.

Now reflect that performing just 7,000 tests can be expected to generate at least 35 false-positive results in total absence of the virus! Obviously, no scientifically sound reasoning underlay any of the plans and measures dictated by the authorities. It cannot be emphasised enough that infection numbers are of no significance if one is not dealing with a truly dangerous virus. Money and means should not be wasted on counting the number of common colds every winter!

Arbitrariness and the lack of a plan wound their way through the measures. At the beginning, facial masks were scorned and not used, even in overcrowded buses. But when the epidemic was over, it became mandatory. DIY stores could stay open for business while electronics markets had to close. Jogging was OK, playing tennis taboo. Every state had its own catalogue of fines; there had to be punishment since we were dealing with an "epidemic of national concern". But where was the logic behind all of these measures? A closer look may help explain what had happened.

The German narrative

Late in the evening of January 27, 2020, the Bavarian Ministry of Health announced Germany's first coronavirus case, an employee of an automotive supplier. A Chinese businesswoman had been on a visit there one week earlier. The virus was subsequently detected in several other members of the company. Most had no symptoms, none was seriously ill. All were isolated and put in a 14-day quarantine. From then on, anyone returning from a "high risk" area, be it China or Tyrol, was tested and put in quarantine. A few scattered numbers of healthy "cases" were thereby discovered.

Then came carnival season in Germany and the western German state of North Rhine-Westphalia is one of its centres where there is no holding back. The first coronavirus patient here had partied in the middle of February together with his wife and 300 other merry carnival revellers in the district of Heinsberg. What happened next sounded the national alarm: coronavirus outbreak in Heinsberg; many patients critically ill; local hospital overwhelmed! Schools and day care centres were closed and all contact persons put in quarantine. At the beginning of March, the Minister of Health, Jens Spahn, still urged prudence. Mass events were cancelled, otherwise overall calmness reigned.

But on March 9, alarm bells rang. The first coronavirus fatalities in Germany occurred. A 78-year old man from the Heinsberg district and an 82-year old woman from Essen succumbed to the virus. The man had a multitude of pre-existing illnesses, among them diabetes and heart disease, the woman died from pneumonia. Drosten warned against a threatening coronavirus wave (115): *"Autumn will be a critical time, that is obvious. At that time, I expect a rapid increase of coronavirus cases with dire consequences and many deaths…Who do we want to save then, a severely ill 80 year-old or a 35 year-old with raging viral pneumonia who would normally die within hours, but would be over the worst after three days on a ventilator?"*.

The pandemic is declared

On March 11, the WHO declared the pandemic. The very next day, German governors of state voted to cancel all mass gatherings. On the same day, a report from France: all day care centres, schools, colleges and universities have been closed until further notice. Germany followed suit: one day later, the German states ordered all schools and day care centres closed from March 16. There was talk of a "tsunami" in the wake of which countless lives would be claimed unless we managed to "flatten the curve". All of a sudden, everyone had a voice and an opinion, no matter whether astrophysicist or trainee journalists, and no matter whether they had not an inkling of knowledge about infectious diseases. Projections were presented every day, exponential growth was explained to us on every channel, showing us how difficult it is to grasp or to even stop this development because the rate of infection seemed to double weekly. Without strict measures we would have one million infections by mid-May. According to RKI President Wieler, the number of fatalities in Germany would soar up and approach Italian numbers within just a few weeks (116).

For the first time, there was mention of a possible lockdown. On March 14, the Federal Ministry of Health tweeted (117):

> » Attention FAKE NEWS!
> It is claimed and rapidly being distributed that the Federal Ministry of Health/Federal government will soon announce further massive restrictions to public life. This is NOT true!

Two days later, on March 16, further massive restrictions to public life were announced (118).

Public life was rapidly shut down. Clubs, museums, trade fairs, cinemas, zoos, everything had to be closed. Religious services were prohibited, playgrounds and sports facilities fenced off. Elective surgery would be postponed. The primary goal: the health care system must not be overwhelmed.

While alarmism was expanding here in Germany, someone else raised his voice. Someone who really knows what he is doing and whom we have heard of several times before, Professor John Ioannidis. Here is a summary of his article "A fiasco in the making?" (119):

The current coronavirus disease, COVID-19, has been called a once-in-a-century pandemic. But it may also be a once-in-a-century evidence fiasco. We lack reliable evidence on how many people have been infected with SARS-CoV-2. Draconian countermeasures have been adopted in many countries. During long-lasting lockdowns, how can policymakers tell if they are doing more good than harm? The data collected so far on how many people are infected and how the epidemic is evolving are utterly unreliable. Given the limited testing to date, some deaths and probably the vast majority of infections due to SARS-CoV-2 are being missed. We don't know if we are failing to capture infections by a factor of three or 300. No countries have reliable data on the prevalence of the virus in a representative random sample of the general population. Reported case fatality rates, like the official 3.4% rate from the World Health Organization, cause horror – and are meaningless. Patients who have been tested for SARS-CoV-2 are disproportionately those with severe symptoms and bad outcomes. The one situation where an entire, closed population was tested was the Diamond Princess cruise ship and its quarantined passengers. The case fatality rate there was 1.0%, but this was a largely elderly population, in which the death rate from COVID-19 is much higher. Adding to these extra sources of uncertainty, reasonable estimates for the case fatality ratio in the general U.S. population vary from 0.05% to 1%. If that is the true rate, locking down the world with potentially tremendous social and financial consequences may be totally irrational. It's like an elephant being attacked by a house cat. Frustrated and trying to avoid the cat, the elephant accidentally jumps off a cliff and dies. Could the

COVID-19 case fatality rate be that low? No, some say, pointing to the high rate in elderly people. However, even some so-called mild or common-cold-type coronaviruses that have been known for decades can have case fatality rates as high as 8% when they infect elderly people in nursing homes. In fact, such "mild" coronaviruses infect tens of millions of people every year, and account for 3% to 11% of those hospitalised in the U.S. with lower respiratory infections each winter. If we had not known about a new virus out there, and had not checked individuals with PCR tests, the number of total deaths due to "influenza-like illness" would not seem unusual this year. At most, we might have casually noted that flu this season seems to be a bit worse than average. The media coverage would have been less than for an NBA game between the two most indifferent teams. One of the bottom lines is that we don't know how long social distancing measures and lockdowns can be maintained without major consequences to the economy, society, and mental health.

Regrettably, this voice of reason remained unheard by our politicians and their advisers. Instead, the prediction ventured by Professor Neil Ferguson, Imperial College London, made the headlines: if nothing is done and the virus allowed to spread uncontrolled, more than 500,000 people will die in the UK and 2 million in the US (120). Not only did this make the rounds, it struck fear into hearts and souls.

Incidentally, Ferguson is the same authority who predicted 136,000 deaths due to mad cow disease (BSE), 200 million deaths due to avian flu and 65,000 deaths during the swine flu – in all cases there were ultimately a few hundred (121). In other words, he was wrong every time. Do journalists actually have a conscience and, if so, why do they not check the facts before distributing their news? Naturally, here too it later became apparent that Ferguson's prediction was totally wrong. But this was never reported by the media.

For the RKI, the headlines seemed to be just the right thing. It warned of an exponential increase (122): "With this exponential growth, the world will have 10 million infections within 100 days if we do not succeed in curbing the number of new infections". Model calculations were published that predicted hundreds of thousands of deaths in Germany (123).

Politicians entered a race for voter popularity – who could profit the most? Markus Söder, State President of Bavaria, presented himself as "Action Man", emanating force and determination in front of the cameras, and declaring his intent to fight the virus to the finish with all the means at his disposal. Söder surges ahead with the first draconian measures: stay-at-home order for Bavarians as of March 21. No visits to loved ones in hospitals. No church services. Shops and restaurants closed. Among other incredible measures.

Nationwide lockdown

What impression would it make on the world if each federal state in Germany had its own rules? So the measures were hastily emulated throughout the nation. The "stay-at-home command" sounded too negative, so we were presented with a "lockdown" on March 23 in the guise of a "nine-point plan". This meant nationwide confinement orders. A far-reaching contact ban was imposed, congregations of more than 2 people in public were forbidden. Restaurants, hair dressers, beauty parlours, massage practices, tattoo studios and similar businesses had to close. Violations of these contact bans were to be monitored by a regulatory agency and failure to comply was to be sanctioned. Penalty catalogues were hastily patched together. Some states went to extremes. Bavaria, Berlin, Brandenburg, the Saarland, Saxony and Saxony-Anhalt enacted decrees that allowed leaving homes and entering public spaces only with a "valid" reason. At the same time, hospitals were so empty that they were able to accommodate patients from Italy and France (124).

On March 25, the German parliament announced an *"epidemic situation of national concern",* so that two days later the hurriedly compiled new "law to protect the population during an epidemic situation of national concern" could be implemented – largely unnoticed by the general population. It empowered the Federal Ministry of Health to determine, by decree, a series of measures that violate the first article of the German constitution: Human dignity is inviolable.

These political decisions were made in the absence of any evidence that might have justified them. It was for that reason that we decided to write an open letter to Chancellor Merkel (28) in which questions of fundamental importance were raised. The intent was to give the government the chance to turn back from the wrong track with dignity. But our opinions, and those of many others who did not agree with the government line, were ignored and dissenting voices were discredited in newspapers and the media. It goes without saying that we never received an answer.

Instead, at the end of March, it was officially proclaimed that the virus was still spreading too fast. Case numbers doubled every 5 days. The goal must be to flatten the curve so the doubling time is extended to 10 days. Only thus would we prevent the health care system from being overwhelmed (125).

The contents of an internal document of the German Ministry of the Interior (GMI) were then released to the public. There one learned that the worst-case scenario forecast 1.15 million fatalities if the virus was not contained (126, 127). If we look at the numbers of reported infections in the first four weeks of March (calendar weeks (CW) 10–13), we can see that this actually looks like exponential growth, exactly as the RKI proclaimed. And that is how it was presented everywhere.

However, what the RKI did not point out was that in calendar week 12 the number of tests had approximately tripled and increased again the following week. The RKI apparently did not feel duty-bound to truth

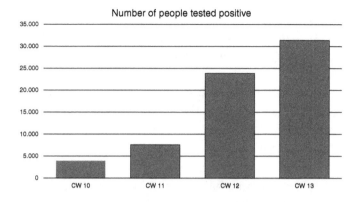

Number of people tested positive

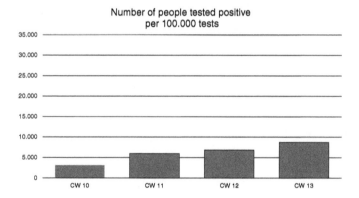

Number of people tested positive
per 100.000 tests

and clarification towards the population. So therefore, are these figures distorted? Why didn't they correct the numbers? That could have been achieved by stating the number of infections per 100,000 tests as shown in the second diagram.

The RKI text should rather have read as follows: "Dear fellow citizens, our numbers show no exponential increase of new infections. There is no need to worry."

Indeed, the epidemic is literally "over the hill", as you can nicely see from the R-curve of the RKI, which was published on April 15 in the Epidemiological Bulletin 17 (128):

What is glaringly evident?

1) The epidemic had reached its peak at the beginning to the middle of March, well before the lockdown on March 23.

2) The lockdown had no effect: numbers dropped no further after its implementation.

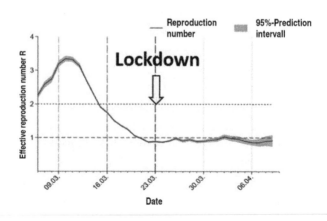

April 2020: no reason to prolong the lockdown

How did things look in the middle of April when the decision of once again prolonging the lockdown was pending?

Everything was really clear now. Just like the R-value, the number of newly infected cases showed that the peak of infection had passed (Figure: www.cidm.online). The upper curve depicts the number of "newly infected" with the initial increase as officially presented; the lower shows those numbers standardized to 100,000 tests. Columns show the actual numbers of conducted tests.

The fact is that there had never been a danger of hospitals being overwhelmed because there had never

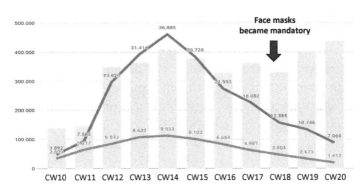

Number of people tested positive

been an exponential growth of infection numbers. There were thousands of empty beds. There never was a giant "wave" of COVID-19 patients. Not because the measures were so effective, but because the epidemic was over before they were put in place. But all the hospitals postponed, or even suspended, all elective surgeries and procedures such as hip or knee operations or check-ups for cancer patients. Many hospitals reported occupancy reductions of up to 30% and more. Doctors were put on short-time working hours (129).

The lockdown is extended

On April 15, Germany extended the lockdown. The rules for social distancing and contact restrictions were prolonged. In public, social distancing of 1.5 m was mandatory and you were only allowed to be outside your domicile with members of your family and one other person who was not part of your household. The ban on meetings in houses of worship was prolonged. Social events were prohibited. Some restrictions were eased. Shops with a retail space of up to 800 square metres were allowed to re-open. Car dealers, bicycle shops and book stores were excluded from this restriction and were allowed to open their doors regardless of size. But amazingly, no matter whether a crocheted scarf or a clinical face mask is used – masks became mandatory!

Mandatory masks

There is simply a lack of clear evidence that people who are not ill or who are not providing care to a patient should wear a mask to reduce influenza or COVID-19 transmission (130).

We are not aware of any single scientifically sound and undisputed article that would contradict the following:

1) There is no scientific evidence that symptom-free people without cough or fever spread the disease.

2) Simple masks do not and cannot stop the virus.

3) Masks do not and cannot protect from infection.

4) Non-medical face masks have very low filter efficiency (131)

5) Cotton surgical masks can be associated with a higher risk of penetration of microorganisms (penetration 97%). Moisture retention, reuse of cloth masks and poor filtration may result in increased risk of infection (132).

Since the government enforced the use of masks, many elderly people believed that they were safe while wearing them. Nothing could be further from the truth. Wearing a mask can entail serious health hazards, especially for people with pulmonary disease and cardiac insufficiency, for patients with anxiety and panic disorders and of course for children. Even the WHO originally stated that general wearing of masks did not serve any purpose (133).

What did the RKI say? In accordance with the shift in political opinion, they also changed their previous recommendations and supported mask-wearing. "If people – even without symptoms – wore masks as a precaution, it could minimize the risk of infection. Of note, this is not scientifically documented."

A report claiming that mask-wearing had provided positive effects was basically flawed (134). According to the study, the effects (drop in numbers of infections) became apparent 3–4 days after implementation of the regulation. However, this is impossible. The RKI states: "An effect of the respective measures can only be seen after a delay of 2–3 weeks because on top of the incubation period (up to 14 days) there is a time delay between illness and receipt of the reports." (135)

In fact, there is no study to even suggest that it makes any sense for healthy individuals to wear masks in public (136, 137). One might suspect that the only political reason for enforcing the measure is to foster fear in the population.

Last argument for extension of lockdown: the impending second wave?

The constant fear-spreading experts of the government obviously pursue the same goal. In Germany, Drosten warned again and again. And somehow it seemed as if every country had its own "Drosten".

At the end of April, he again fantasized about the big-time wave in Germany – now, of course, the second big wave (138): "Would the R-value through carelessness … be once again more than 1 and thereby exponentially increase virus spread, this would likely have devastating consequences. Since the wave of infection would start everywhere at the same time, it would have a different momentum."

But where should this second wave of infection come from?

Drosten: We can learn this from the Spanish flu. It started at the end of the First World War, and most of the 50 million victims died during the second wave.

That is true. But at the time of the Spanish flu, antibiotics were not available to treat secondary bacterial infections that were the main cause of death (139). Consequently, people of all ages died. Whoever compares COVID-19 to the Spanish flu is either completely clueless or deliberately intends to spread fear.

It is clear that viruses change but do not simply disappear. Just as there has always been a flu season, there has also always been a coronavirus season (140).

Here we see the typical course of a coronavirus epidemic (141):

Does this look vaguely familiar and reminiscent of our RKI data with the March peak?

But wait, this Finnish study stems from 1998!

So, if any government should decide they want a second wave, all they need to do is to radically increase the number of tests in the annual coronavirus season. This simple manipulation will not fail to trigger the next laboratory pandemic.

Relaxing the restrictions with the emergency brake applied

Professor Stefan Homburg, Director of the Institute of Public Finance at the University of Hannover, never tired of explaining why the RKI numbers themselves called for immediate termination of all measures (142).

He was not the only one, several others raised their

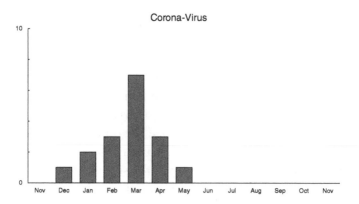

Corona-Virus

voices. But critical opinions were completely ignored. Why? Did the government have an exclusive contract with Drosten, who keeps on warning and warning: by loosening restrictions, Germany will risk losing its lead in the fight against the pandemic (143).

But eventually the time arrived. The beginning of May witnessed a cautious reopening of shops. Schools and day care centres would soon be able to admit children again. Contact restrictions were slightly relaxed and life was restarted, but at a painfully slow pace.

But the RKI warns and warns and warns (144): "The reproduction factor is more than 1 once again. It's at 1.1, to be exact ... ".

Horror of horrors, were we too rash? Many were puzzled that the daily R-factor fluctuated erratically. This of course was due to the generally unknown fact that when infection numbers are very low, the R-factor can be manipulated at will simply by altering the number of tests conducted.

And then, the great scare: Do we possibly have excess mortality (145)?

Excess mortality? Really? Could it possibly have anything to do with the collateral damage invoked by the unwarranted measures? This question was posed by a senior member of the risk analysis division at the German Ministry of the Interior. He produced a remarkable document in which the risks of collateral damage were meticulously analysed. He arrived at the conclusion that the measures were excessive, and that they caused immense and irreparable collateral damage without providing any true benefits. The synopsis of the paper was sent to ten external experts, including ourselves, to have the numbers checked.

He then attempted to present the document to the Minister: unsuccessfully. He then sent the document to his colleagues in risk assessment divisions around the country. And was suspended for his efforts.

We stated in a press release that we considered the conclusions of the paper to be very important. But the

Ministry ridiculed the document, saying that it was no more than a private opinion (146). The media chimed in and considered the case closed.

Lockdown extended again!

At the end of May, just before the agreement on contact restrictions between the government and the federal states expired, a further extension of the measures was proclaimed until June 29.

On May 25, Minister of Health, Jens Spahn stated in the most widespread German daily newspaper, "Under no circumstances should the impression be gained that the pandemic is already over."

Only chancellor Merkel could top this – and so she did 4 days later. In an historic declaration, she announces to the depressed nation: "The pandemic has just begun!"

And this at a time when the epidemics were all over throughout Europe.

But an extension of the lockdown seemed to make sense in the light of a recent article published in *Nature,* one of the most prestigious scientific journals in the world. Only research groups of high standing have realistic chances of seeing their names in print in this journal. Imperial College London rallied such a group, among whom the name Neil Ferguson may ring a bell. In a remarkable study, the investigators presented a computer-based analysis showing that the global lockdown had saved many millions of lives (147).

Known only to few was the fact that a string of protests by scientists of international standing rained into *Nature's* office. All pointed to the fundamental flaws in the analysis that had caused false conclusions to be drawn. Correctly handled, the data actually showed the opposite: the lockdown had had no effect on the course of the pandemic. Readers who wish to read the paper should not forget to look at these critical comments that follow after the article (148).

So, while other countries like Denmark at no time recommended that healthy people who move around in public generally wear face masks (149) and other countries like Latvia were well on their way to freedom, Merkel and friends decided against too much liberty for their people. The masks must stay on!

Too much? Too little? What happened?

Overburdened hospitals

The pictures from Italy and Spain incited fear. Mortally ill people and no available ventilators? How dreadful. Deaths were depicted as slow, merciless drownings. We were shown what happens when hospital capacity reaches its limits and beyond. During all the deliberations about what was to be done in Germany, there was always – first and foremost – the fear stoked by the RKI that such scenarios happening in Germany could not be ruled out. As a result, ventilators were purchased, intensive care beds were held in reserve, operations were postponed or cancelled. In Berlin a new hospital for 1,000 patients was hurriedly built – in 38 days – and then, when it was completed, not one patient in sight (150).

We simply must take a closer look at this. At the beginning of March it became clear that the epidemic was sweeping through Germany. Was our health care system well prepared? Professor Uwe Janssens, President of the Interdisciplinary Association of Intensive Care and Emergency Medicine, gave the all-clear in the "Deutschlandfunk" (German World Service) (151): *"We have enough intensive care beds!"*. Even if we were to have as many coronavirus infections as Italy, we had approximately 28,000 beds in intensive care units, 25,000 of which were equipped with ventilators, so nearly 34 beds per 100,000 citizens. This was like no other country in Europe. Professor Reinhard Busse, leader of the specialist field "Management of the Health Care System" at the Technical University in Berlin, gave the all-clear as well: *"Even if we had conditions like in Italy, we would be nowhere near to being overburdened"* (152).

But the RKI kept fostering fear. The "number of intensive care beds will not be sufficient", Wieler, president of the RKI and trained veterinarian, announced at the beginning of April (153). Why? Wieler explained: *"The epidemic continues and the number of fatalities will keep going up"*.

Actually, the real explanation – kept under lock and key at that time – was quite different. It came to light in May, when a previously confidential document appeared on the website of the German Ministry of the Interior (154). The shocking contents confirmed circulating rumours. The document, dating to mid-March, was the minutes of a meeting of the coronavirus task-force. There, one was astounded to learn that fear-mongering was the official agenda created to manage the epidemic. All the pieces of the puzzle then fell into place. Everything had been planned. The high numbers of infection were purposely reported because the numbers of deaths would "sound too trivial". The central goal was to achieve a massive shock effect. Three examples are given how to stir up primal fears in the general population:

1) People should be scared by a detailed description of dying from COVID-19 as "slow drowning". Imagining death through excruciating slow suffocation incites the most dread.

2) People should be told that children were a dangerous source of infection because they would unwittingly carry the deadly virus and kill their parents.

3) Warnings about alarming late consequences of SARS-CoV-2 infections were to be scattered. Even though not formally proven to exist, they would frighten people.

Altogether, this strategy would enable all intended measures to be implemented with general acceptance by the public.

HORRIBLE!

Now that the method in the madness is known, it becomes more understandable why Wieler steadfastly adhered to his projections. Numbers of infections were used to calculate the number of intensive care beds that would be needed, without taking into account that 90% of infected individuals would not fall seriously ill. And that the majority of patients who did require hospitalisation would recover and be dismissed.

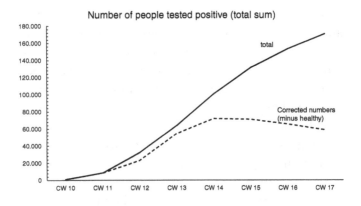

Number of people tested positive (total sum)

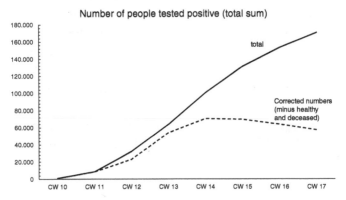

Number of people tested positive (total sum)

Simply adding the daily number of new infections to the curve (top curves in the graph) was of course senseless. The recoveries should have been subtracted from the number of positively tested persons if a realistic indicator of hospital burden had really been sought.

Strictly speaking, one would also have to subtract the deceased, but since there were so few – tragic and sad as that was for every individual case, it made no difference in the graphic representation.

The fact is that we were never at any risk of our health care system collapsing. In mid-April there was NO REASON for further measures. All should have been revoked immediately. While the hospitals waited for non-existent coronavirus patients, those genuinely requiring treatment were not admitted. Beds were empty. Hospitals ran into financial problems. Many applied for short-time work for doctors and nursing staff – in the midst of the imagined crisis (155). The situation in other countries was similar. Thousands of US physicians were placed on administrative leave because the number of routine outpatient visits dropped by a landslide (156).

Shortage of ventilators?

At the commencement of the pandemic, experts contended that invasive ventilation would be a first-line requirement to rescue COVID-19 patients from a horrible death by suffocation. At the same time, this measure would minimize the risk of infection of medical personnel. As a consequence, the German government decided to purchase and store thousands of ventilators in reserve.

This turned out to be a very bad bet (157–161).

Artificially ventilated patients require very close attention (162). Oxygen is forced through a tube into the lungs. It is not uncommon for bacteria to hitch a ride and then cause life-threatening pneumonia. The risk of these hospital-acquired infections rises by the day, which is why medical students learn that the ventilator should be used no longer than is absolutely necessary.

In contrast, COVID-19 patients were often put on ventilation early and without true need, and kept on the apparatus far longer than they ever should have been. Why? Because it was officially stipulated that invasive ventilation was the best means to reduce the risk of virus spread via aerosol to the personnel. However, aerosols probably play no important role in disease transmission (163). The sole fact that SARS-CoV-2 can be found in aerosol droplets (164) does not mean that it is there in sufficient quantities to cause illness (165).

How many lives were lost because of this advice?

Many specialists later stated that COVID-19 patients were intubated and ventilated for too long and too often (160, 161). The risks were high and success more than questionable. Professor Gerhard Laier-Groeneveld from the lung clinic in Neustadt advised that intubation should be avoided in any event. His COVID-19 patients received oxygen with simple respiratory masks and he lost not a single life (160).

Professor Thomas Voshaar, Chair of the Association of Pneumology Clinics, shared the same view (161). He pointed out that the high death rates in other countries "should be reason enough to question this strategy of early intubation". At the time of his report, he had mechanically ventilated one of his 40 patients. The patient subsequently died. All the others survived.

Here is a shortened version of a radio interview with

palliative physician Dr Matthias Thöns (166): "Politics these days has a very one-sided orientation towards intensive care treatment, towards buying more ventilators and offering ICU beds as a reward. But we must remember that most of the severely ill COVID-19 patients are very old people with multiple underlying diseases; 40% of those come heavily care-dependent from assisted living facilities. Previously, this group would ordinarily receive more palliative instead of intensive care. But now, a new disease is diagnosed and this whole client base is turned into intensive care patients."

He points out that according to a Chinese study, 97% die despite maximal therapy (including ventilation). Of those who survive, only a small number is able to return to their former lives, many of them left with severe disabilities. These are circumstances that most seniors would refuse to risk. He rightly says that critically ill patients should openly be told the truth about their condition. They should themselves decide which course they wish to take: intensive care treatment in isolation, or symptomatic treatment in the circle of loved ones. The individual will should have highest priority. Thöns is quite sure that most people would prefer the second option.

Were the measures appropriate?

It became clear fairly early that SARS-CoV-2 was not a killer virus and there never had been an exponential increase in new infections. The price for attempting to contain the virus was absurdly high.

What did the government do right?

- ?

The authors have no answer to this question. They look forward to receiving yours.

What did the government do wrong?

- It proclaimed an epidemic of national concern that did not exist
- It deprived citizens of their rights
- It made arbitrary instead of evidence-based decisions
- It intentionally spread fear
- It enforced senseless lockdown and mask-wearing
- It devastated the economy and destroyed livelihoods
- It disrupted the health care system
- It inflicted immense suffering on the populace

What should our government have done?

It should have done what the chancellor and ministers solemnly declared when they were sworn into office:

"I swear that I will use my power for the WELL-BEING of the German public, to further its ADVAN-TAGES, to prevent DAMAGE, to PRESERVE and DEFEND the constitution and the federal statutes, to diligently fulfil my duty and practice just treatment to-wards everyone."

Collateral damage

Dr David L. Katz, President of the True Health Initiative, asked on March 20 if our fight against the coronavirus was worse than the disease (167). Could there not be more specific means to combat the disease? What about all the collateral damage?

Stanford Professor Scott Atlas said during an interview that under the misassumption that we have to contain COVID-19, we have created a catastrophic situation in the health care sector (168). Irrational fears were generated because the disease as a whole is a mild one. Thus, there is no reason for comprehensive testing in the general population and it should be done only where appropriate, namely in hospitals and nursing homes. At the end of April, Atlas published an article entitled "The data are in – stop the panic and total isolation" (169).

In Germany, Wolfgang Schäuble, presiding officer of the German parliament, stated that not absolutely everything must be subordinate to the protection of life (170).

"If there is anything at all that has an absolute value in our constitution, it is human dignity which is inviolable. But it does not preclude that we have to die."

The media immediately flared back in righteous disgust: "Human dignity versus human life – can you balance one against the other?" (171).

Many still fail to comprehend that we have sacrificed both.

Proponents of the pointless measures argue that every person has the right to grow as old as possible. Even if the virus were only the straw that broke the camel's back, it was still at fault. Without the virus, the deceased may have lived months or even years longer. It is our moral duty to sacrifice our personal wants and needs when lives of others are at stake. The economy can recover, the dead cannot. The Merkel mantra, chanted day and night by her ardent followers: "Protecting the health of our citizens must, at all costs, remain our supreme goal."

Honourable as this may sound, it betrays an alarming inability to comprehend the essence of public welfare. The following numbers have already been presented but because of their importance, they will be repeated here. During the course of this entire epidemic, a maximum number of 10 in 10,000 over 80 year-olds have died with or from the virus. The number of "true" COVID-19 deaths cannot be higher than 1–2 per 10,000. How many human lives were really prolonged by the horrendous measures? Maybe 2–4 per 10,000? Or even 4–8? But definitely not more. And at what cost?

The one employee of the GMI who dared to compile an analysis of the collateral damage to the health care system was suspended. The government was not interested. Nothing can be placed over human life. But what are the consequences for health and welfare of the populace if the economy collapses and people are confronted with the end of their existence?

Economic consequences

It will strike all countries. The global economic crisis could plunge 500 million people into poverty, so stated in a position paper by the UN (172).

The US Federal Reserve (FED) expects a dramatic decline of up to 30% in American economic performance (173). FED director Jerome Powell assumes a 20% to 25% increase in the unemployment rate. Almost 36.5 million people have lost their jobs. It is "the most traumatic job loss in the history of the US economy," says Gregory Daco, US Chief Economist of the Oxford Economics Institute (174).

The EU commission predicts a deep recession of historic magnitude for Europe (175).

According to their prognosis, the economy will shrink a good 7% and will not completely recover in the next year.

In Germany too, the economy is starting to crumble. Since the second half of March it is down to 80% of normal economic performance (176). Reduced hours compensation is registered for about 10 million employees. Without short-time work, the unemployment rate would have increased dramatically, similar to the US.

In April we have "only" 300,000 additional unemployed (177). But this will not be the end of the story, not by a long shot.

The government boasted that they are weaving safety nets, the "greatest rescue package in Germany's history" will help mitigate the collateral damage (178). But that rescue package is ridiculous in relation to the damage that has been done. Countless people are falling through the net. Existences have been destroyed and lives have been lost. They cannot be salvaged by safety nets.

Disruption of medical care

- Many who were ill were afraid to visit hospitals for fear of catching the "killer virus".
- - Often older people would rather not "be a burden" to their doctors, who they thought were battling to save COVID-19 patients.
- Patients requiring medical examinations were turned away, all that was not deemed of "vital importance" cancelled or postponed.
- Medical check-ups were not performed.
- Operations were postponed to free up capacity for "coronavirus patients".
- Domestic violence against women and children increased.
- The number of suicides rose.

Drugs and suicide

Following the financial crisis of 2008, the number of suicides rose in countries all over the world. According to the National Health Group Well Being Trust, unemployment, economic downfall and despair could now drive 75,000 Americans to drug abuse and suicide (179). The Australian government estimates a rise in suicides of 50% (180), a number 10 times higher than the number of "coronavirus deaths". Unemployment and poverty are also predicted to markedly increase suicide rates in Germany (181).

Heart attack and stroke

Unemployment increases the risk of heart attack to an extent comparable to cigarette smoking, diabetes and hypertension (182). But where did all the patients with heart attacks disappear to? Admissions to emergency care units dropped 30% as compared to the previous month. Not because the patients were miraculously cured but because they were terrified of catching the deadly virus in the hospital. Preliminary symptoms went unheeded, even though such symptoms are often the harbinger of a deadly attack and need to be closely attended to in hospital.

"This is a most dangerous development... There are now 50% fewer patients with mild symptoms in the emergency room," explains Dr Sven Thonke, chief

physician at the Clinic for Neurology in Hanau in a newspaper interview (181). Many pending strokes initially cause mild symptoms such as dizziness, speech, visual problems and muscle weakness. Thonke: "There are now 50% fewer patients with mild-symptoms in the emergency room." This is extremely worrisome because more often than not mild symptoms herald the severe stroke that can be rapidly fatal if the emergency is not immediately tended to.

Other ailments

According to the scientific institute of the AOK (German health insurance company), the following diagnoses dropped considerably in April: 51% fewer respiratory diseases, 47% fewer diseases of the digestive tract, and 29% fewer injuries and poisonings (183).

Care of tumour patients was catastrophic. Monitoring of tumour treatment was no longer conducted at the required levels. Control examinations were postponed or cancelled. Patients waited in agony for the next appointment – alone with their fears and the single remaining question: how much time was still left to them.

Cancelled operations

30 million elective surgeries were postponed or cancelled worldwide during the first 12 weeks of the pandemic (184). In 2018, 1.4 million operations were performed on average every month. 50–90% of all scheduled operations were postponed or not performed in March, April and May 2020. This translates to at least 2 million operations that would normally have been performed. The consequences must be profound.

Further consequences for the elderly

In Germany, more than 1,000 people over the age of 80 die every day (185). While we are taking drastic measures to prevent them from dying of COVID-19, we are making their lives less worth living. This cannot but impinge on life expectancy.

Quality of life

Especially in old age – when many friends have already passed on and the body no longer works the way it once did – life is not about how many more days or years but about a life worth living. That could be accomplished by exercise and remaining active, through social contacts, by taking recreational holidays, visiting events and even shopping sprees, with regular visits to the sauna or a fitness studio or the daily walk to the corner café.

But what happens when, all of a sudden, the café and everything else is closed? No more visits to old friends, no more social events. And no visitors either.

Loneliness and isolation

Functioning social networks safeguard the elderly from loneliness. Five to twenty percent of senior German citizens feel lonely and isolated. After the lockdown, almost all contact with other people stopped for months, which must have worsened these feelings. For those who cannot leave the house unassisted, nursing services arrange "senior social groups", where the elderly are picked up once a week and then taken safely home again. It's not much, but it's so important to be with other people again and devastating when no longer there.

Terminal care

Yes, every individual has the right to reach as old an age as possible. But every person nearing the end of their life should also have the right to decide how they want to go. Most do not fear the end. As the time approaches, people become increasingly detached and willing to embark on their last journey.

When we hear talk about the "older people" and we are told that it is our moral duty to protect them, many picture sprightly seniors who are enjoying their time on ocean liners. In reality, the endangered elderly are multi-morbid individuals at the end of their lives. People who have not been able to leave their beds for days, weeks or months. People whose tumours have spread throughout their bodies and are in constant pain. People who cannot go on anymore and maybe do not want to go on. People who sometimes just wait for a kind fate to relieve them of their suffering.

Amidst all the protective measures for the high-risk groups in retirement and nursing homes, at the end the individual decision should have the highest priority. Most no longer care whether their loved ones bring the coronavirus to them, as long as someone is there to hold their hand, to talk about the past, and to whisper I love you and farewell (186).

Innocent and vulnerable: our children

Children – like the elderly – are the most vulnerable in our society and it is our duty to care for them. Millions of children in the world are suffering acutely from the coronavirus measures. "The coronavirus strikes more children and their families than those who are actually gripped by the infections," says Cornelius Williams, Head of the UNICEF Child Protection League (187).

Mental/psychological stress

Children cannot thrive without social contacts. Separation from key people like grandma and grandpa, auntie and uncle, their best friends – the closed schools, inaccessible playgrounds and barred sports fields disrupt their lives. Social ethicists point out how vital it is for children to be in contact with their peers (188).

Educational deficits

Children have a right to education. Since the schools have been closed, millions of students are lagging behind according to an estimate of the German Teacher Association. Their president, Heinz-Peter Meidinger, sees educational deficits for approximately 3 million children, especially in students from difficult social backgrounds and from impoverished families (189).

Physical violence

Tens of thousands of children in Germany become victims of violence and abuse every year (190). Crime statistics from 2018 show that

- 3 children die in the aftermath of physical violence every week
- 10 children are physically or mentally abused every day
- 40 children are sexually abused every day

And these, of course, are only the known cases. Can you imagine the situation in coronavirus times?

- When parents are stressed, on the brink of losing their jobs and facing financial ruin?
- When arguments and quarrels become a daily occurrence?
- With increased alcohol consumption?
- When children are at home day after day, with no way of escape?

Teachers who normally play important roles in safeguarding endangered children are gone. Who then should notify the youth welfare office should the need arise?

The government's commissioner for abuse, Johannes-Wilhelm Rörig, issued an urgent warning. There were indications from the quarantined town of Wuhan that the cases of domestic violence had tripled during the "trapped-at-home" time. There were "equally alarming numbers" from Italy and Spain.

Consequences for the world's poorest

Many in this country took the opportunity to get their house and garden back into shape during the coronavirus crisis. Understandably, since home-office work was only semi-effective for want of equipment and slow internet connections. Actually, the majority of the middle class and the affluent were not doing badly. Well, the neighbour who now has to apply for Hartz IV (unemployment benefits) will surely get back on his feet. People tend to think as far as their front door, maybe a bit beyond, but that's it. Many are not aware that the most severe consequences often affect the poorest of the poor. One must not close one's eyes to the fact that the existence and lives of countless people are threatened.

Existential consequences

In India, there are hundreds of millions of day-labourers, many of whom led a hand-to-mouth existence before the coronavirus restrictions robbed them of their livelihoods. Now they have no more means to survive. They are "protected" against the coronavirus and are in turn left to starve.

In many African countries, coronavirus lockdowns are brutally enforced by police and military. Whoever shows his face on the streets is beaten. Children, who usually survive on their one meal in school, are forbidden to leave the house. They, too, can starve.

At the end of April, the Head of the UN World Food Program, David Beasley, gave a warning before the UN Security Council: because of coronavirus, there is a danger that the world will face a "hunger pandemic of biblical proportions" (191). "It is expected that lockdowns and economic recessions will lead to a drastic loss of income among the working poor. On top of this, financial aid from overseas will decrease, which will hit countries like Haiti, Nepal and Somalia, just to name a few. Loss of revenue from tourism will doom countries like Ethiopia, since it represents 47 percent of national income."

Consequences for medical care and maintenance of health

Medical care is a luxury that only a few in the poorest countries can afford. Advances and positive developments of recent years are now in danger of collapse.

Vaccination campaigns against the measles were suspended in many countries. Although measles rarely cause death in western countries, 3–6% of the infected people in poor countries die, and those who survive often have life-long disabilities. The virus has claimed 6,500 child deaths in the Congo Republic (192).

Between 2003 and 2013, Zimbabwe succeeded in lowering yearly malaria infections from 155 per 1,000 inhabitants to just 22. Now, and within a short time, there have been more than 130 deaths and 135,000 infections. Two thirds of all fatalities were < 5 year-old children.

According to the WHO, malaria deaths in sub-Saharan Africa could rise to 769,000 in 2020, which would double the number for 2018. If so, they would be thrown back to a "mortality standard" of 20 years ago. The probable reason for this catastrophe is the fact that insecticide-treated mosquito nets can no longer be adequately distributed.

Are the malaria deaths in Zimbabwe and the measles deaths in the Congo only precursors of what is in store for the continent?

Synopsis

With the prescribed measures, was our government able to prolong the lives of people who would leave us in the next days, months or perhaps a few years? Maybe, maybe not. Were many lives saved through these measures? They certainly were not, because these restrictions were imposed when the epidemic was already subsiding.

One thing is certain. The immeasurable grief that these measures have inflicted cannot possibly be put into words or numbers.

Did other countries fare better – Sweden as a role model?

While we were lectured every day on the "pseudo-exponential" growth of infections and talked into thinking that our health system would collapse if drastic measures were not strictly enforced, a few other countries chose a different path. They did not establish a curfew, they left restaurants, fitness studios, and libraries etc. open to the public. Sweden is an example (193).

Epidemiologist Professor Anders Tegnell, who obviously learned from mistakes he had made during the swine-flu epidemic, and his predecessor, Johan Giesecke, who at an early stage pointed out that only the implementation of evidence-based measures made any sense, both decided that lockdowns were not only pointless, but dangerous. Giesecke explained in an interview (194):

"There are only two measures that have a genuine scientific background. One of these is hand-washing and we know this since the work of Ignaz Semmelweis 150 years ago. The other is social distancing. Many of the measures taken by European governments have no

scientific basis. Closing the borders for example is use-less and does not help. Also, the closing of schools has never proven to be effective."

From a scientific stance, school closings are indeed known to make no sense (89).

It did make sense, however, to count on the individual sense of responsibility of the citizens, and on informational and educational campaigns. People were informed on how to protect themselves – and they did: without fear-mongering, without panic scenarios, without lockdown, without threat of a fine, without massive restrictions on their liberties.

Executive WHO director Mike Ryan called Sweden a "role model" in the fight against the coronavirus (195).

Undeniably, Sweden did a lot of things right. But it reaped disgust and disapproval from its neighbours. The German press left no stone unturned to badmouth the Swedish way:

- Sweden's special path apparently failed (Deutschlandfunk, April 4, 2020)
- Consequences cannot be predicted – 10% mortality rate: Sweden's lax special path during the coronavirus crisis is threatening to fail (Focus, April 17, 2020)
- Coronavirus in Sweden – Is the country heading for a catastrophe? (RND, April 24, 2020)

Politicians also had their say.

Karl Lauterbach (SPD) accused Swedish men and women of acting irresponsibly. "Crudely put, many of the elderly are sacrificed so that the cafés do not have to close."

Minister-President of Bavaria, Markus Söder, said: "This liberal course claims VERY, VERY MANY victims …"

As a matter of fact, the epidemic in Sweden took a comparable course as that in other countries.

Homburg describes this in an interview (196): "It seems that they want to avoid at all costs acknowledging that there is an example to the opposite of their own misguided policy. They have tried with every means at their disposal – fake news followed by more fake news – to throw Sweden off its chosen path. But Sweden stayed the course."

Could we have taken this path in Germany? Count on the individual sense of responsibility of the citizens and on information campaigns?

A favourite counter argument is Sweden's population density. With 23 inhabitants per square kilometre it is about 10 times lower than in Germany, so it is argued that it might work there, but never here. This would also apply to Iceland, which is another positive example of how to master the coronavirus crisis without lockdowns. Almost all of the 1,800 infected people recovered. 10 COVID-19 deaths were registered – without any drastic lockdown. Many restaurants and schools remained open and congregations of up to 20 people were allowed.

This may be true, but here we also have a low population density. So let us look instead at Hong Kong with 7.5 million residents and a population density of 6,890 people per square kilometre. And what a surprise: Here, too, it worked! It was a little more restrictive than Sweden and Iceland maybe, but nevertheless without complete lockdown (197).

Or let us look at Japan (126 million inhabitants, population density 336 per square kilometre) or South Korea.

Japan and South Korea were among the first countries outside of China to be affected by the outbreak. Contrary to China's draconian measures, the mass quarantines in wide parts of Europe and in major US cities, regular life continued in Japan for a large part of the population. Restaurants stayed open – without a serious disaster (198). Japan has a very small number of coronavirus infections – possibly because they did not do much testing.

Now, we know that the number of infections is of no significance. So let us look at the really important issue, namely the number of deceased: this, too, is extremely low. Much wrong cannot have been done in Japan!

In contrast to Japan, South Korea performed more testing than any other country, but shutdown of public life was also largely avoided. No cities were cordoned off, nor general curfews imposed. Public institutions, shops, restaurants and cafés stayed open (199).

South Korea banked on 1) informing the public and 2) testing and tracing. Mass testing was performed in specially erected drive-through centres. Radical transparency was ensured by a tracking app that tagged the whereabouts of the infected persons.

Sweden, Iceland, Hong Kong, South Korea, Japan – all these examples have confirmed what recognised experts have said all along: lockdowns are not necessary. They cause massive social and economic damage that cannot justify any possible benefits. But were there benefits at all?

Are there benefits of lockdown measures?

At the end of 2019, the WHO published a document describing various measures to be taken in case of a future pandemic (200). The major goal would be, as we have heard before, to "flatten the curve" by reducing the number of new daily infections. A number of measures were considered "Out" from the very beginning: they were NOT recommended IN ANY CIRCUM-STANCES!

Hmm – so how come everything happened as it did? If it had been possible, would the world have also been put under UV-light and the humidity raised beyond the tropics?

After telling us what should definitely not be done, the WHO went on to describe other measures – lockdown etc. – that it deemed more worthy of recommendation. Hidden in an appendix was, admittedly, a note that the recommendations had no scientific basis.

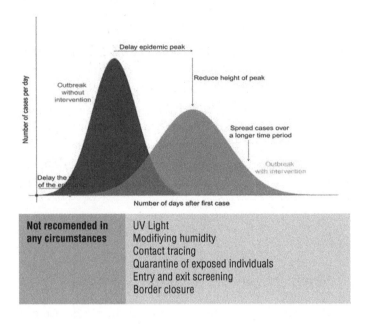

Not recomended in any circumstances	UV Light Modifiying humidity Contact tracing Quarantine of exposed individuals Entry and exit screening Border closure

Several critical scientists came to the conclusion early on that lockdown was the wrong path. Among others, Nobel laureate Professor Michael Levitt spoke out. He considered the lockdown a "gigantic mistake" and called for more appropriate measures that should specifically aim to protect the vulnerable groups (201).

Nonetheless, most countries followed the "role model" China.

All of Italy was completely quarantined from March 10 by a stay-at-home order. Exceptions applied only in emergencies, for important work orders and for errands that could not be postponed. 60 million people were under house arrest and the streets were totally empty for a whole two months. Other countries like Spain, France, Ireland, Poland undertook similar action. With what effect? The epidemic is over, so let us look at the death toll – keeping in mind that the numbers are grossly inflated because of faulty counting methods and case definition.

Did fewer people die in countries with lockdown measures?

When we look at the death rates per 1 million inhabitants for some European countries with lockdown (alphabetically, first 13 columns), we see that the numbers appear to vary quite considerably. The median number is around 340 (red bar represents mean with standard deviation). Realise, however, that this is low in comparison to something in the order of 10,000 deaths per million that occur annually in Germany and other European countries. And that the coronavirus numbers are grossly exaggerated because most derive from deaths with rather than death from the virus. Divide them by at least 5 to arrive at realistic numbers. Then, the variations lose meaning. Respiratory infections caused by many agents similarly sweep like gusts of wind that blow 20 or 100 of 10,000 leaves from a tree. Every loss is sad, but most are fateful. Preventive measures need to be appropriate so as to avoid collateral damage that would sweep other leaves from the tree.

The press relentlessly emphasized that Sweden would pay a high price for its liberal path. In actuality,

Number of deaths (per 1 million population)

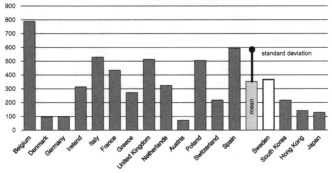

we see that Sweden without lockdown is not significantly different when compared to countries with lockdown. South Korea, Japan and Hong Kong as well do not conspicuously stand out with an exorbitantly high number of so called "corona deaths". Quite the contrary is the case.

So what do we see: countries without lockdown measures did not slide into a catastrophe.

We know that COVID-19 can run a fatal course in elderly patients with underlying conditions. This leads to the next important question.

Were high-risk groups better protected in countries with lockdown?

The simple answer is, No.

94

Approximately half of the "coronavirus victims" died in care facilities and retirement homes, no matter where you look. In Western countries, these numbers vary from 30% to 60% (202). Countries with relatively drastic lockdowns like Ireland (60%), Norway (60%) or France (51%) have no better figures than Sweden (45%). Nursing homes require specific protection which general lockdown measures can in no way achieve.

A sensible concept for protection of genuinely vulnerable groups compliant with ethical rules and regulations (203) would have solved the problem.

Would immediate suspension of the lockdown have had dire consequences?

Let us look at the Czech Republic. From March 16, curfews were instated, citizens were only allowed to go to work, to go grocery shopping, to see a doctor or to go for walks in public parks. Like everywhere, the lockdown could not prevent the increase in infections. By court decision, the measures had to be rescinded on April 24. Was there a new wave of new infections and deadly casualties? Oh – it really seems so! Is the Czech Republic experiencing the much-feared second wave of COVID-19 infections – a scenario feared all across the continent? Of course not! The number of tests has been increased (204).

These data just illustrate how irrelevant and misleading the numbers of false-positive "new cases" are

Daily New Cases in Czechia

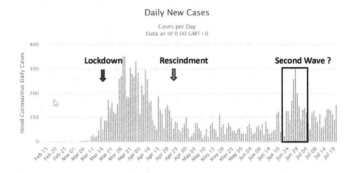

Daily New Cases

Cases per Day
Data as of 0.00 GMT+0

Daily New Deaths in Czechia

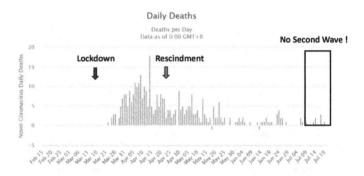

Daily Deaths

Deaths per Day
Data as of 0.00 GMT+8

when the virus is more or less gone. This is confirmed by looking at the number of daily deaths. With a corresponding delay due to the incubation period, there should be a significant increase in the middle of July (rectangle). But the numbers kept sinking and the epidemic in the country was over as well (Worldometers, July 2020).

This scenario of another "wave of infections" is typical for many countries. It is often misused to maintain fear in the population and to prolong senseless measures (205).

In fact, the epidemic followed essentially the same course all over Europe. The effects of the lockdown were exclusively negative.

In a few countries such as Israel, there currently seems to be a second increase in the number of daily deaths. Media revel in spreading news of the dreaded second wave. But do not be fooled. Look closely and inform yourself. Numbers must always be set in relation – to the number of residents, number of PCR tests, average number of total deaths. If the number of people who die with a positive SARS-CoV-2 PCR test is small, as in Israel, perfectly irrelevant increases (e.g. from 2 to 6) can be turned into sensational news: the death toll has tripled! Interestingly, at the height of the COVID-19 epidemic in March, Israel's overall deaths per month dropped to the lowest rate in four years. So there was never even a first "COVID-19 wave". In July, the number of so-called "COVID-19 deaths" per 1 million population was not even half as high as in Germany (Worldometers, July 2020).

So which measures would have actually been correct?

Simple: a resolute protection of the vulnerable groups, especially those in nursing and care facilities. Period.

Is vaccination the universal remedy?

»There can be no return to normality until we have a vaccine," declares Michael Kretschmer, Minister-President of Saxony (206).

More and more voices were raised that we needed a vaccine before we could return to normal life.

At the beginning of June, the German Federal Ministry of Finance issued a plan to boost the economy: Item 53: "The coronavirus pandemic ends when a vaccine is available" (207)! This is hysterical! Since when can a government decide how and when a pandemic ends?

On Easter Sunday, Bill Gates was allotted ten minutes prime time to address the German nation on television (208).

Ingo Zamperoni (TV- host): "It is becoming increasingly clear that we can only get a grip on this pandemic if we develop a vaccine."

Bill Gates: "We will ultimately administer this newly developed vaccine to 7 billion people, so we cannot afford problems with adverse side effects. However, we will make the decision to use the vaccine on a smaller data basis than usual. This will enable rapid progress to be made."

Rapid progress on a small data basis? Is this the right way to fight a disease with relative low fatality rate?

Remarkably, start-up financing for the global search for a coronavirus vaccine was accomplished at the beginning of May by sleight of hand. The EU collected almost 7.5 billion euro with their donor conference. Germany and France pledged a large portion. A special programme was launched by our government to serve this purpose. The plan is to contribute 750 million euro toward the development of a vaccine.

But does vaccination really make sense? How vulnerable are we towards the virus? How many lives are threatened that need to be protected?

On the question of immunity against COVID-19

A short excursion into the field of immunology.

What does immunity against coronaviruses depend on?

The coronavirus binds via protein projections (so-called spikes) that recognise specific molecules (receptors) on our cell. This can be likened to virus hands grasping the handles of doors that then open to allow entry. After multiplication, viral progenies are released and can infect other cells.

Immunity against coronaviruses rests on two pillars: 1) antibodies, 2) specialised cells of our immune system, the so-called helper lymphocytes and killer lymphocytes.

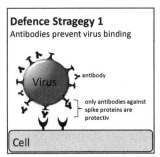

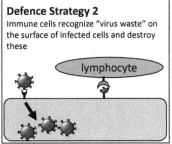

When a new virus enters the body and causes illness, the immune system responds by mobilising these arms of defence. Both are trained to specifically recognise the invading virus, and both are endowed with the gift of long-term memory. Upon re-invasion by the virus, they are recruited to the new battle sites, their prowess bolstered through their previous encounter with the sparring partner.

Many different antibodies are generated, each specifically recognising a tiny part of the virus. Note that only the antibodies that bind the "hands" of the virus are protective because they can stop the virus from gripping the handles of the door (step 1). Classical viral vaccines are designed to make our immune system produce such antibodies. It is believed that an individual will thus become immune to the virus.

Three points require emphasis.

  1. If you are tested for SARS-CoV-2 antibodies and nothing is found this does not mean that you were not infected. Severe symptoms often correlate with high production of antibodies, mild symptoms only lead to low antibody levels and many asymptomatic infections probably occur without any antibody production.

  2. If antibodies are found this does not mean that you are immune. Current immunological tests cannot selectively detect protective antibodies directed against the "hands" of the virus. Other antibodies show up at the same time. Testing cannot give any reliable information on the "im-

mune status" of an individual and, as will follow next, is essentially useless.

3. The outcome of an encounter between "protective" antibodies and the virus is not "black or white", not a "now or never". Numbers are important. A wall of protecting antibodies may ward off a small attack – for instance when someone coughs at a distance. The attack intensifies as the person comes closer. The scales begin to tip. Some viruses may now overcome the barrier and make it into the cells. If the cough comes from close quarters, the battle becomes one-sided and ends in a quick victory for the virus.

So even if vaccination is "successful", meaning that production of protective antibodies has taken place, it does not guarantee immunity. To worsen matters, antibody production spontaneously wanes after just a few months. Protection, if any at all, is at best short-lived.

The idea of a personal "Immune Status" document is scientifically unsound.

What happens after the virus enters the cell? Experiments conducted on mice have examined this in detail for SARS-CoV, the original SARS virus and close relative of the present SARS-CoV-2. It was demonstrated that the second arm of the immune system comes into play. Lymphocytes arrive on the scene. A coordinated series of events takes place during which helper cells explode into action and activate their partners, the killer lymphocytes (209). These seek out the cells that contain the virus and kill them. The factory is destroyed – the fire is extinguished.

Cough and fever go away.

How can killer lymphocytes know which cells to attack? Put in simple words: imagine an infected cell to be a factory that produces and assembles the virus parts. Bits and pieces that are not assembled into the viruses become waste products that the cell removes in an ingenious way: it transports them out and puts them in front of the door. The patrolling killer cells see the trash and move in for the kill (step 2).

This second arm of our immune system is seldom talked about, but it is probably actually all- important – much more so than the antibodies that represent a rather shaky first line of defence. Most importantly, waste products derived from different coronaviruses share similarities. Killer lymphocytes recognising the waste of one virus can therefore be expected to recognise at least some of the waste of others.

Would this imply cross-immunity?

Yes. Coronavirus mutations take place in very small steps. Protective antibodies and lymphocytes against type A will therefore also be quite effective against progeny Aa. If B comes to visit, you get another cold and cough, but then your immune status broadens to cover A, Aa, B and Bb.

The scope of immunity expands with each new infection. And lymphocytes can remember.

Who does not recall their child's first year in kindergarten? Oh no, not again, here comes the umpteenth cold with runny nose, cough and fever. The child is ill all through the long winter! Luckily, it gets better the second year and the third will be weathered with maybe just one or two colds. By the time school starts, the operational base for combating the viruses has grown rock solid.

So what does "Immunity against coronavirus" really mean?

Does "immune" mean that we do not get infected at all?

No. It means *we don't fall seriously ill.*

And not getting sick does not rest solely on prevention of infection by antibodies, but more on "putting out the fire". When a new variant appears, many people may get infected but because the fires are quickly extinguished, they will not fall seriously ill. The relative few who fare worse do so because the balance between attack and defence is heavily in favour of the virus. But in the absence of pre-existing illness, the scales tip back again. The virus will be overcome. As a rule, it is only for people with pre-existing conditions that the virus may become the last straw that breaks the proverbial camel's back.

This is why coronavirus infections run a mild or

even symptom-free course and why an epidemic with any "new" virus is never followed by a second, more serious, wave.

Why do annual coronavirus epidemics end in summer? Well, just one speculation. Over 50% of the northern European population becomes vitamin D-deficient in the dark winter months. Possibly, replenishment of vitamin D stores by sunshine and the shift of activities to outdoors are simple important reasons.

What happens to the virus after an epidemic? It joins its relatives and circulates with them in the population. Infections continue to occur but most go unnoticed because of the vitalised immune system. Once in a while, someone will get his summer flu. But such is life.

Can a similar pattern be expected with SARS-CoV-2?

The authors believe that is exactly what we have witnessed. 85–90% of the SARS-CoV-2 positive individuals did not fall ill. Most probably, their lymphocytes extinguished the fires in time to limit viral production. Put very simply: the virus was a new variant and able to infect almost anyone. But immunity was already widespread due to the presence of lymphocytes that cross-recognised the virus.

Does proof exist that lymphocytes from unexposed individuals cross-recognise SARS-CoV-2?

Yes. In a recent German study, lymphocytes from 185 blood samples obtained between 2007 and 2019 were examined for cross-recognition of SARS-CoV-2. Positive results were found in no less than 70–80%, and this applied to both helper and killer lymphocytes (210). A US study with lymphocytes from 20 unexposed donors similarly reported the presence of lymphocytes that were cross-reactive with the new virus (211). In these and another Swedish study it was also found that even non-symptomatic or mild SARS-CoV-2 infections provoked strong T-cell responses (212). We suspect that these unusually vigorous T- cell responses to a first infection represent classical booster phenomena occurring in pre-existing populations of reactive T-lymphocytes.

Could the idea that lymphocytes mediate cross-immunity to SARS-CoV-2 be tested?

The concept of lymphocyte-mediated herd immunity that we present follows from the integration of latest scientific data (209–212) into the established context of host immunity to viral infections. The idea can actually be put to test. Thus, in a recent study, cynomolgus monkeys were successfully infected with SARS-CoV-2 (213). Although all animals shed the virus, not a single one fell ill. Minor lesions were found in the lungs of two animals, attesting to the fact that vigorous production of the virus had taken place.

In essence, these findings replicated what has been witnessed in healthy humans. Repetition of the monkey experiment in animals depleted of lymphocytes would show whether herd immunity had indeed derived from the presence of the cells.

To vaccinate or not to vaccinate, that is the question

The development of vaccines against dreaded diseases such as smallpox, diphtheria, tetanus and poliomyelitis represented turning points in the history of medicine. Vaccination against a number of further diseases followed which today belong to the standard repertoire of preventive medicine. Now, the most pressing issue arises whether a global vaccination programme is needed to end the coronavirus crisis. This question is so important that a debate urgently needs to be conducted to reach a global consensus on three basic points.

1. When is the development of a vaccine called for? We venture to answer: when an infection regularly leads to severe illness and/or serious sequelae in healthy individuals, as is not the case with SARS-CoV-2.

2. When would mass vaccination not be reasonable? We propose that mass vaccination is not reasonable if a large part of the population is already sufficiently protected against life-threatening disease, as is the case for SARS-CoV-2.

3. When will vaccination likely be unsuccessful? We predict that vaccination will fail when a virus co-existing worldwide with man and animals continuously undergoes mutational change, and when individuals become exposed to high doses of virus during spread of the infection.

In the authors' view, a global vaccination programme thus makes no sense. The risks far outweigh any possible benefit right from the start.

Experts around the world express their concerns and warn of rushed COVID-19 vaccines without sufficient safety guarantees (214, 215).

Yet, researchers are currently working on more than 150 COVID-19 vaccine candidates (216), with some already in advanced clinical trials. The aim of most vaccines is to achieve high levels of neutralising antibodies against the binding spike proteins of the virus and cellular responses (217, 218). Four major strategies are being followed.

1. **Inactivated or attenuated whole virus vaccines.** Inactivated vaccines require production of large quantities of the virus, which need to be grown in chicken eggs or in immortalised cell lines. There is always the risk that a virus batch will contain dangerous contaminants and produce severe side effects. Moreover, the possibility exists that vaccination may actually worsen the course of subsequent infection (219), as has been observed in the past with inactivated measles and respiratory syncytial virus vaccine (220,221).

Attenuated vaccines contain replicating viruses that have lost their ability to cause disease. The classic example was the oral polio vaccine that was in use for decades before tragic outbreaks of polio occurred in Africa that were found to be caused not by wild virus, but by the oral vaccine (222).

2. **Protein vaccines**. These will contain the virus spike protein or fragments thereof. Supplementation with immune stimulators, adjuvants that may cause serious side-effects, is always necessary (217).

3. **Viral vectors as gene-based vaccines.** The principle here is to integrate the relevant coronavirus gene into the gene of a carrier virus (e.g. adenovirus) that infects our cells (217). Replication-defective vectors are unable to amplify their genome and will deliver just one copy of the vaccine gene into the cell. To bolster effectiveness, attempts have been made to create replication-competent vaccines. This was undertaken with the Ebola vaccine rVSV-ZEBOV. However, viral multiplication caused severe side effects in at least 20% of the vaccinated, including rash, vasculitis, dermatitis and arthralgia.

4. **Gene-based vaccines.** In these cases, the viral gene is delivered to the cell either as DNA inserted into a plasmid or as mRNA that is directly translated into protein following cell uptake.
A great potential danger of DNA-based vaccines is the integration of plasmid DNA into the cell

genome (223). Insertional mutagenesis occurs rarely but can become a realistic danger when the number of events is very large, i.e. as in mass vaccination of a population. If insertion occurs in cells of the reproductive system, the altered genetic information will be transmitted from mother to child. Other dangers of DNA vaccines are production of anti-DNA antibodies and autoimmune reactions (224).

Safety concerns linked to mRNA vaccines include systemic inflammation and potential toxic effects (225).

A further immense danger looms that applies equally to mRNA-based coronavirus vaccines. At some time during or after production of the viral spike, waste products of the protein must be expected to become exposed on the surface of targeted cells. The majority of healthy individuals have killer lymphocytes that recognise these viral products (210,211). It is inevitable that autoimmune attacks will be mounted against the cells. Where, when, and with which effects this might occur is entirely unknown. But the prospects are simply terrifying.

Yet, hundreds of volunteers who were never informed of these unavoidable risks have already received injections of DNA and mRNA vaccines encoding the spike protein of the virus, and many more are soon to follow. No gene-based vaccine has even received approval for human use, and the present coronavirus vaccines have not undergone preclinical testing as normally required by international regulations. Germany, a country whose populace widely rejects genetic manipulation of food and opposes animal experiments, now stands at the forefront of these genetic experiments on humans. Laws and safety regulations have been bypassed in a manner that would, under normal circumstances, never be possible. Is this perhaps why the government still declares an "epidemic situation of national concern" to exist – in the absence of serious new infections? For then the new German Infection Protection Act empowers the government to make exceptions to the provisions of the Medicinal Products Act, the medical device regulations, and regulations on occupational safety and health. And this has given the green light to the fast-track vaccine development project.

But the authors wonder whether the Infection Protection Act can go so far as to permit genetic experiments be conducted on humans who have not been informed of the potential dangers.

Pandemic or no pandemic – the role of the WHO

Actually, have we not had a lighter version of pandemic-driven vaccination hype before?

Exactly the same thing happened with the "swine flu" in 2009. Everyone was told that a vaccine was desperately needed to stop the deadly pandemic. Vaccines were then produced at miraculous speed – and sold *en masse* to states around the world.

Prior to 2009, a pandemic required three criteria to be met (226):

- The pathogen must be new
- The pathogen must spread and cross continents rapidly
- The pathogen must generally cause serious and often fatal disease

The swine flu turned out to meet the first two criteria, but not the third. Because the call to declare a pandemic was very pressing, especially from the pharmaceutical industry, major financers of the WHO (227), the WHO cut the Gordian knot with a stroke of genius. A pandemic, it declared, can take a mild or serious course!

In 2010, the definition of a pandemic was simplified yet further as "the worldwide spread of a new disease". Flu and coronaviruses continuously undergo mutation and it is to be expected that variants will occasionally emerge that cause somewhat atypical disease that could then be dubbed as "new". The swine flu provided the stage for a first exercise in the employment of panic-making strategy to handle a pandemic. A typical headline: *"Swine flu: the calm before the storm?"* (228) appeared in December of 2009 when it was clear that virtually no one was ill and the course of the infection had been milder than previous waves of influenza. Still, virologists warned of underestimating the "dangerous" virus: *"If we look at this virus in an animal experiment and compare it with preceding viruses, one sees that the virus is not harmless at all! It is much more dangerous than the annual H3N2-virus."*

Brilliant. But what does this have to do with human medicine? Which prominent scientist spread this frightening conclusion with such conviction? Ah yes, a certain Professor Drosten.

The article continues: When, in the coming Christmas days, the Germans vigorously intermix their viruses, a second wave seems inevitable. This could be considerably more severe than the first.

A second wave was predicted, with the medical health system being hopelessly overwhelmed, says, not Professor Drosten for once, but Professor Peters from the University of Münster. He feared that the number of beds in intensive care units would be insufficient. Moreover, many patients would need artificial respiration. Dramatic situations could be created in the overwhelmed hospitals.

Are you also having déjà-vu right now?

A nationwide vaccination with the hastily produced and barely tested H1N1 vaccine was recommended. 60 million doses of adjuvanted vaccine were purchased for the German population. Non-adjuvanted vaccine was obtained only for high members of the government (229).

Again, this all happened when it was clear that the swine flu pandemic had run a light course. The majority of the public decided wisely against the senseless vaccination. What was the end of the story? Trucks loaded with over 50 million expired vaccine doses were disposed of at the Magdeburg waste-to-energy plant. As was taxpayer's money ... no, actually not, the money just changed hands. Estimated profit for the pharmaceutical industry: 18 billion US dollars (230).

Actually, that was not quite the end of the fiasco. Almost forgotten today is that one adjuvanted swine flu vaccine caused side effects that ruined thousands of lives (231,232). The side effects were caused because antibodies against the virus cross-reacted with a target in the brains of the victims. The damage was the result of a classic antibody-driven autoimmune disease. The side-effect was relatively rare. The incidence was probably something in the order of 1 in 10,000, but the outcome was tragic because so many millions received the vaccine, essentially for nothing, since the infection generally ran a mild course. In retrospect, the risk-benefit ratio of swine flu vaccination must be admitted to have been disastrous. This is what happens when mass vaccination is undertaken without need.

Failure of the public media

It's easier to fool people than to convince them that they have been fooled.
(MARK TWAIN)

In a working democracy, the media should provide the public with truthful news, foster opinion formation through critique and discussion, and oversee the action of the government as the "fourth public authority" with impartiality and autonomy. What we have experienced during the coronavirus pandemic is just the opposite (233).

All public broadcasters became servile mouthpieces of the government. The press was no better. Regard for the truth, protection of human dignity, service to the public – the Press Codex disappeared from the scene. Worldwide.

Where was truthful information to be found?

And where were critical discussions of any information?

We were presented with disturbing pictures and frightening numbers – morning, noon and night. Someone was always issuing a warning somewhere – Drosten, Wieler, Spahn, Merkel. No one in the media ever critically questioned these warnings or investigated their truth.

Scaring the population seemed to be the sole agenda (234). Reports on millions of fatal casualties were presented without mention that they were based on model calculations. No mention was made that Ferguson, the producer of these numbers, had always been completely wrong in his numerous doom-forecasting predictions.

At the same time, the media abstained from questioning how the RKI numbers were compiled, what they meant and what could, or rather could not, be gathered from them. Instead, the figures were uncritically accepted and used to unsettle the public.

Where was the open discussion?

It could hardly have been more monotonous. Always the same "experts" – of which there were apparently only two in Germany. Why was there never a discussion between the government advisers and the critics like Dr Wolfgang Wodarg, a lung specialist and board member of the anti-corruption organisation "Transparency International" Germany? An open and objective exchange: Drosten and Wieler and Bhakdi and Wodarg together at a round-table talk. Well, it did not hinge on Bhakdi or Wodarg or many other critics of the government course. It was simply not wanted by the government.

There was much talk about how the Swedish way without lockdown was being criticised by Swedish experts. That the German way was also massively criticised by many knowledgeable citizens in their own country was never a subject of discussion.

Besides Wodarg, the immunologist and toxicologist Professor Stefan Hockertz pointed out early on that the seriousness of SARS-CoV-2 should be assessed similar to that of the common flu viruses, and that the implemented measures were completely exaggerated. Also involved was Christof Kuhbandner, a professor of psychology, who reiterated several times that there was no scientific basis for these measures (235). How could he know, people asked? The interesting thing is that any observant person with a fundamental understanding of number theory can take the time to analyse the statistics and come to the same conclusion. There are topics that span across multiple disciplines. Dr Bodo Schiffmann, an ear-nose-and throat specialist from Sinsheim, did the job that the journalists should have done. Almost daily he posted videos on his YouTube channel with indefatigable energy and persistence to inform the public on the latest developments and to explain the numbers and why they were wrong.

The critical voices in this country were not alone, there were many others worldwide (236, 237). Was the public notified? It seemed to have been an easy and successful strategy to simply not report these things; but such a stratagem should have no place in an enlightened democratic state.

This synchronised "system journalism" was obviously apparent to experts. Professor Otfried Jarren voiced his criticism in the Deutschlandfunk (238). *"For weeks now, the same male and female experts and politicians make their appearance and are presented as the "crisis managers". But nobody asks who has which expertise and who appears in which role. Furthermore, there are no debates among these experts, but only individual statements."*

The numbers game

You can do a lot with numbers. Above all, you can make people afraid.

Example 1: infection rate. The infection rate was continuously increasing, soon our health care system would collapse – what they didn't say was that the number of recovered people was also continuously increasing and that there were no grounds for such an assumption. That remained a secret.

Example 2: mortality rate. "The US had the highest number of deaths worldwide." On May 28, the nightly news reports showed images of the deceased: *"They all died from COVID-19. With more than 100,000 deaths, the US is mourning the highest number of victims worldwide."* Now we know that a big fraction of these poor people did not die from COVID-19, but rather from the measures taken against COVID-19.

Also, the US is the third largest country in the world. So perhaps it would make more sense to look at the number of deaths per 100,000 inhabitants? This number was relatively low – very much below the numbers from Spain or Italy. Was that not worth mentioning? Furthermore, a good journalist could also point out that the "number of deaths" is not an absolute value, not the least because the counting methods are different for every country.

The country with the highest mortality rate per 100,000 citizens was Belgium. The numbers were much higher than in Spain or Italy. Was the situation there really so dramatic? No. As already shown, the basic problem related to the method of counting (45). If such facts are not reported by the media, then of course the numbers cannot be correctly assessed.

Defamation and discrediting

When critical voices were heard, immediate action was taken to silence them by defamation. The lung specialist Wolfgang Wodarg was the first to raise his voice. The defamation campaign that followed was unparalleled.

As soon as we had published our first YouTube videos warning about the excessive measures and pointed out that Italy might have other aggravating factors, e.g. the high levels of air pollution), there was the first "facts-check". Under the headline "*Why Sucharit Bhakdi's numbers are wrong*", an article was quickly put into the "ZDF Mediathek". Nils Metzger supposedly gets to the bottom if this (239): "*Biology professor downplays coronavirus danger*". A good starting point since the title immediately suggested that we were not dealing with a medical doctor who had seen countless patients and was a specialist in infection epidemiology, but with a biologist. And at some point the classic situation whereby things are put into your mouth that you have never said – just to discredit you. Metzger: "*To present the factor air pollution as the sole trigger for the crisis – as Sucharit Bhakdi did in his video – is unscientific.*" Naturally it was never once claimed anywhere that the high number of victims was solely due to air pollution, because that would indeed have been unscientific. This statement was a blatant lie. But ARD/ZDF believers would hardly have made the effort to check the "real" facts. Unfortunately, there are still a lot of people who think that things must be true when they are reported by the public broadcasters. Sadly, that is not the case.

Censorship of opinions

Article 5 of the German constitution:

Article 5 [Freedom of expression]
(1) Every person shall have the right freely to express and disseminate his opinions in speech, writing, and pictures and to inform himself without hindrance from generally accessible sources. Freedom of the press and freedom of reporting by means of broadcasts and films shall be guaranteed. There shall be no censorship.

There is no place for critical opinions in either the public press or the public broadcasts. The only alternative was by means of the social media, where the public could be informed via YouTube videos. But even here, freedom of expression is merely lip service. You can find quite a few videos that get away unpunished even though they promote lies, hate and agitation. YouTube apparently has no problem with those. However, an interview with the Austrian TV station Servus TV about coronavirus was deleted. This happened to a lot of videos that were critically involved in this topic. Susan Wojcicki, CEO of YouTube, said during an interview (240): *"Everything that violates the recommendations of the WHO would constitute a breach against our guidelines. Therefore, deletion is another important part of our guidelines."* The WHO that was responsible for the fake swine flu pandemic in 2009; The WHO that overestimated the COVID-19 mortality on a large scale, and drove the world into a crisis with this and other misjudgements?

This same WHO that now sets the standard on what can be said?

WhatsApp reacted as well. The forward function was restricted in order to contain the distribution of Fake News during the coronavirus crisis. But who exactly determines if news is fake? What if our own government distributes Fake News? On March 14, the Ministry of Health warned via Twitter:

>> Attention FAKE NEWS!
It is claimed and rapidly distributed that the Federal Ministry of Health/Federal government will soon announce further massive restrictions to public life. This is NOT true!

Two days later, on March 16, further massive restrictions to public life were announced.

The English Professor John Oxford, one of the best-known virologists worldwide, said the following about the coronavirus crisis (241): *"Personally, I would say the best advice is to spend less time watching TV news which is sensational and not very good. Personally, I view this COVID outbreak as akin to a bad winter influenza epidemic. We are suffering from a media epidemic!"*

The German "good citizen" and the failure of politics

It is easier to believe a lie that you have heard a thousand times than to believe a truth that you have only heard once (ABRAHAM LINCOLN)

We had a division within the country once before – during the refugee issue. The opinions varied widely and there was talk about "good citizens", the do-gooders and "angry citizens", the not so do-gooders.

This time it is a lot worse. Friendships break apart. People face each other with irreconcilable differences. They talk about each other, against each other – but not with each other. Some are driven by worries about collateral damages; others see themselves as advocates for the rights of the elderly who are to be sacrificed for the economy.

Here is a commentary from a local paper after Chancellor Angela Merkel addressed the nation with the decision to extend the lockdown:

"I was very relieved. Relieved, that we apparently did everything right with our social distancing, our sacrifice by not meeting friends or visiting family and all of that. I was very relieved that we will continue this in the future". Sadly, this is not an individual opinion. The media epidemic claimed a lot of victims.

Eminent psychologist, Professor Gerd Gigerenzer, addressed this issue (234):

"It is easy to trigger a fear of shock risks in people. These are situations where a lot of people die suddenly in a very short time. This new coronavirus could be such a shock risk, just the same as plane crashes, acts of terror or other pandemics. If, however, deaths are spread out over a year, it hardly scares us even if the number is significantly higher."

Indeed. Without any measures having had any effect at all and at the end of the epidemic, we are looking at far fewer than 10,000 so called "coronavirus deaths" in Germany (Worldometers, July 2020).

In Germany, approximately 950,000 people die each year. Of those, more than a third (350,000) die of cardiovascular diseases and 230,000 of cancer (242).

Many of these 950,000 deaths could be prevented by information and education, starting in schools and continuing for the general public, about the importance of exercise and healthy diets, about the dangers of obesity and many other issues. We could prevent thousands of deaths each year. And we might also have fewer deaths from respiratory diseases, whereby a small virus would not break the camel's back, because that back would not be strained to the breaking point. This applies not only to the coronaviruses but to many other viruses and bacteria that have always done that and will continue to do so in the future.

Why did our politicians fail?

After he had understood everything, a colleague exclaimed: *"But how can that be? It either means that our government and their advisers are completely ignorant or incompetent – or, if they are not, there MUST be some kind of intention behind it. How else can you possibly explain all this?"*

Helmut Schmidt, Chancellor of the Federal Republic of Germany from 1974 to 1982, was one of the last German politicians with class. He once said: *"The stupidity of governments should never be underestimated."* He was right, of course, but THIS stupid? Really? One cannot and does not want to believe that. Therefore, that only leaves the second question – what is the intention behind all of this? And now politicians are wondering why "conspiracy theorists" are springing up like mushrooms. Why did our government ignore other opinions and make decisions haphazardly and without a solid basis? Why did our government not act in the general interest and for the good of the German people?

According to Johann Giesecke, politicians wanted to use the pandemic to advance their own positions and were perfectly willing to implement measures that were not scientifically substantiated (196). *"Politicians want to demonstrate their capacity to act, the capacity for decision making and most of all their strength. My best example for this is that in Asian countries the sidewalks are sprayed with chlorine. This is completely useless but it shows that the state and the authorities are doing something, and that is very important to politicians."* There are some indications from Austria that he could be right in this:

During their crisis management, the Austrian government did not trust in the expertise of their own advisers. An interview transcript later revealed that Chancellor Sebastian Kurz was counting on fears rather than explanations when implementing the rigid measures, which made it easier to get the public to accept social and economic impositions (243).

The strategy document of the German Ministry of the Interior reveals that the same agenda had been premeditated in this country (154).

Why was there so little criticism of the government's course from the economy?

The stock market professional, Dirk Müller, gave a persuasive explanation why the pandemic was a blessing for many (244): in short, because it is always the same story: Big companies win, small ones lose. Big corporations will survive while many small and midsize companies as well as private businesses will perish. Finance professor, Stefan Homburg, called it *"the largest redistribution of wealth in peacetime"*. The loser would be the taxpayer (245).

Why was there so little criticism from the scientists' ranks?

Let's not be naïve. Science is just as corrupt as politics. The European Union provided 10 million euro for research on the novel coronavirus. Every Tom, Dick and Harry who wanted to research this virus could apply for financing. So very soon now we will have a lot of, possibly useless, information about SARS-CoV-2 and under these circumstances it is not exactly helpful to point out the relative harmlessness of the virus.

Conclusions:

- the government is committed to serving the good of the citizens
- the opposition is committed to oversee government action
- the press is committed to inform the public by critical and truthful reporting
- those in the know (in this case physicians and scientists) are obligated to raise their voice and demand evidence-based decisions

Every citizen who did not attend to his duties is an accomplice to the collateral damage of the coronavirus crisis.

Quo vadis?

You can fool all the people some of the time, and some of the people all the time, but you cannot fool all the people all the time (ABRAHAM LINCOLN)

The relevant authorities, our politicians and their advisers played truly inglorious roles in the handling of new and supposedly dangerous infections of the last decades, from BSE, swine flu, EHEC to COVID-19. At no point did they learn from their mistakes, and this diminishes the hope that it will be any different in the future. On the contrary! While we "only" redistributed taxpayers' money to the pharmaceutical industry during the swine flu, this time livelihoods were destroyed, the constitution was trampled on and the population basically deprived of their fundamental rights: freedom of speech and opinion, freedom of movement, freedom of relocation, freedom of assembly, freedom of actively practicing your religion, freedom to practice your occupation and make a living.

Anchored in the constitution is the principle of

proportionality: the State's interference with basic rights must be appropriate to reach the aspired goal. And last but not least: the dignity of mankind must never be violated.

This ceased to be the case, to the detriment of democracy and civilisation.

It has been almost 90 years since the time when critical and free journalism was abolished and the media transformed into the extended arm of the state.

It has been almost 90 years since the time when freedom was abolished and opinions of the public were forced into the political line.

It has been almost 90 years since the last media-driven mass hysteria.

If we have learned just one thing from the darkest times of our German history, then surely this: We must never again be indifferent and look the other way. Especially not when the government suspends our fundamental democratic rights. This time, it was only a virus that knocked on our door, but look what we had to go through as a consequence:

- Media-fuelled mass hysteria
- Arbitrary political decisions
- Massive restrictions of fundamental rights
- Censorship of freedom of expression
- Enforced conformity of the media
- Defamation of dissidents (the differently minded)
- Denunciation
- Dangerous human experiments

If that does not remind you of a dictatorship then you must have been sound asleep during your history lessons. The things that remain with us are deep concern and fear. Because so many intelligent and educated people became like lemmings within a short 3 months, willing to obey the demands and commands of the world elite.

The renowned virologist Pablo Goldschmidt said (246): *"We are all locked up. In Nice there are drones that impose fines on people. How far has this monitoring gotten? You have to read Hannah Arendt and look very closely at the origins of totalitarianism at that time. If you scare the population, you can do anything with it."*

Apparently, he is right. One thing is clear: there are many things that should be worked through and we should all insist upon this happening. The coronaviruses have retreated for this season, the issue is disappearing from the headlines and from the public sphere – and soon it will be gone from peoples' memories.

If we, the people, do not demand that all transgressions of the coronavirus politics are addressed, then those in power will be able to cover it all with a cloak of concealment.

There is always the chance of some other threat knocking on our door. The only positive thing that has come from this is that very many people in our country have woken up. Many have realised that the mainstream media and politicians can agree to support each other on things that are not good – and even evil. One can only hope that the admonishing voices of reason will in future not be silenced by the dark forces on this earth.

A farewell

Respiratory viruses are a major cause of mortality worldwide, with an estimated 2–3 annual million deaths. Many viruses including influenza A viruses, rhinoviruses, respiratory syncytial virus (RSV), parainfluenza viruses, adenoviruses and coronaviruses are responsible. Now, a new member has joined the list. As with the others, the SARS-CoV-2 virus particularly endangers the elderly with serious pre-existing conditions. Depending on the country and region, 0.02 to 0.4% of these infections are fatal, which is comparable to a seasonal flu. SARS-CoV-2 therefore must not be assigned any special significance as a respiratory pathogen.

The SARS-CoV-2 outbreak was never an epidemic of national concern. Implementing the exceptional regulations of the Infection Protection Act were and still are unfounded. In mid-April 2020, it was entirely evident that the epidemic was coming to an end and that the inappropriate preventive measures were causing irreparable collateral damage in all walks of life. Yet, the government continues its destructive crusade against the spook virus, thereby utterly disregarding the fundaments of true democracy.

And as you read these lines, human experiments are underway with gene-based vaccines whose ominous dangers have never been revealed to the thousands of unknowing volunteers.

We are bearing witness to the downfall and destruction of our heritage, to the end of the age of enlightenment.

May this little book awaken *homo sapiens* of this earth to rise and live up to their name. And put an end to this senseless self-destruction.

References

(1)   "Coronavirus Disease (COVID-19) Weekly Epidemiological
        Update and Weekly Operational Update," World Health
        Organization, last accessed August 26, 2020, https://www.who.int
        /emergencies/diseases/novel-coronavirus-2019/situation-reports.
(2)   Chih-Cheng Lai et al., "Severe Acute Respiratory Syndrome
        Coronavirus 2 (SARS-CoV-2) and Coronavirus Disease-2019
        (COVID-19): The Epidemic and the Challenges," *International
        Journal of Antimicrobial Agents* 55, no. 3 (March 2020):
        105924, https://doi.org/10.1016/j.ijantimicag.2020.105924.
(3)   Catrin Sohrabi et al., "World Health Organization Declares
        Global Emergency: A Review of the 2019 Novel Coronavirus
        (COVID-19)," *International Journal of Surgery* 76 (April 2020):
        71–76, https://doi.org/10.1016/j.ijsu.2020.02.034.
(4)   Shuo Su et al., "Epidemiology, Genetic Recombination, and
        Pathogenesis of Coronaviruses," *Trends in Microbiology* 24, no.
        6 (June 2016): 490–502, https://doi.org/10.1016/j.tim.2016
        .03.003.
(5)   Jie Cui, Fang Li, and Zheng-Li Shi, "Origin and Evolution of
        Pathogenic Coronaviruses," *Nature Reviews Microbiology* 17
        (2019): 181–92, https://doi.org/10.1038/s41579-018-0118-9.
(6)   Yanis Roussel et al., "SARS-CoV-2: Fear Versus Data,"
        International Journal of Antimicrobial Agents 55, no. 5 (May
        2020): 105947, https://doi.org/10.1016/j.ijantimicag
        .2020.105947.
(7)   David M. Patrick et al., "An Outbreak of Human Coronavirus
        OC43 Infection and Serological Cross-Reactivity with SARS
        Coronavirus," *Canadian Journal of Infectious Diseases and
        Medical Microbiology* 17, no. 6 (November–December 2006):
        330–36, https://www.ncbi.nlm.nih.gov/pmc/articles
        /PMC2095096.
(8)   Kelvin K. W. To et al., "From SARS to Coronavirus to Novel
        Animal and Human Coronaviruses," *Journal of Thoracic Disease*
        5, no. S2 (August 2013): S103–8, http://doi.org/10.3978/j.issn
        .2072-1439.2013.06.02.

(9)   "SARS (Severe Acute Respiratory Syndrome)," National Health
        Service (UK), last reviewed October 24, 2019, https://www.nhs
        .uk/conditions/sars.
(10)   "Middle East Respiratory Syndrome Coronavirus (MERS-CoV),"
        World Health Organization, https://www.who.int/emergencies
        /mers-cov/en.
(11)   "COVID-19 Coronavirus Pandemic," Worldometer, https://www
        .worldometers.info/coronavirus.
(12)   Benjamin Reuter, "Coronavirus lässt in Italien Ärzte
        verzweifeln—Entscheidungen wie in Kriegszeiten," *Der
        Tagesspiegel* (Berlin), March 12, 2020, https://www.tagesspiegel
        .de/wissen/drohen-in-deutschlanditalienische-verhaeltnisse
        -coronavirus-laesst-in-italienaerzteverzweifeln-entscheidungen
        -wie-in-kriegszeiten/25632790.html.
(13)   Ciro Indolfi and Carmen Spaccarotella, "The Outbreak of
        COVID-19 in Italy: Fighting the Pandemic," *JACC: Case Reports*
        2, no. 9 (July 2020): 1414–18, https://doi.org/10.1016/j.jaccas
        .2020.03.012.
(14)   Max Roser et al., "Coronavirus Pandemic (COVID-19)," *Our
        World in Data*, last updated August 26, 2020, https://ourworld
        indata.org/mortality-risk-covid.
(15)   "Coronavirus Disease 2019 (COVID-19): Situation Report—61,"
        World Health Organization, March 20, 2020, https://www
        .who.int/docs/default-source/coronaviruse/situation-reports
        /20200321-sitrep-61-covid-19.pdf.
(16)   Michael Day, "COVID-19: Four Fifths of Cases Are
        Asymptomatic, China Figures Indicate," *BMJ* 369, (April 2020):
        m1375, https://doi.org/10.1136/bmj.m1375.
(17)   "Regeln zur Durchführung der ärztlichen Leichenschau," *AWMF
        Online* (Germany), revised January to October 2017, https://
        www.awmf.org/uploads/tx_szleitlinien/054-002l_S1_Regeln-zur
        -Durchfuehrung-der-aerztlichen-Leichenschau_2018-02_01.pdf.
(18)   Audrey Giraud-Gatineau et al., "Comparison of Mortality
        Associated with Respiratory Viral Infections between December
        2019 and March 2020 with That of the Previous Year in
        Southeastern France," *International Journal of Infectious
        Diseases* 96 (July 2020): 154–56, https://doi.org/10.1016/j.ijid
        .2020.05.001.
(19)   Victor M. Corman et al., "Detection of 2019 Novel Coronavirus
        (2019-nCoV) by Real-Time RT-PCR," *Eurosurveillance* 25, no. 3
        (January 2020): 2000045, https://doi.org/10.2807/1560-7917
        .ES.2020.25.3.2000045.
(20)   Sonja Gurris, "Corona-Tests werden Geheimwaffe," *n-tv*
        (Cologne), March 30, 2020, https://www.n-tv.de/panorama
        /Corona-Tests-werden-Geheimwaffe-article21678629.html.

(21) Christian Drosten, Twitter post, April 13, 2020, 4:42 p.m., https://twitter.com/c_drosten/status/1249800091164192771.

(22) Australian Associated Press, "WHO Rejects Tanzania Claim Tests Faulty," *Examiner* (Launceston), May 8, 2020, https://www.examiner.com.au/story/6749732/who-rejects-tanzania-claim-tests-faulty.

(23) Yafang Li et al., "Stability Issues of RT-PCR Testing of SARS-CoV-2 for Hospitalized Patients Clinically Diagnosed with COVID-19," *Journal of Medical Virology* 92, no. 7 (July 2020): 903–8, https://doi.org/10.1002/jmv.25786.

(24) Gurris, "Corona-Tests werden Geheimwaffe."

(25) Ines Nastali, "Police Intervenes on Quarantined Mein Schiff 3," *Safety at Sea*, May 6, 2020, https://safetyatsea.net/news/2020/police-intervenes-on-quarantined-mein-schiff-3-2.

(26) "Wenig Infektionen beim Charité-Personal," *Deutsches Ärzteblatt* (Berlin), May 13, 2020, https://www.aerzteblatt.de/nachrichten/112809/Wenig-Infektionen-beim-Charite-Personal.

(27) John P. A. Ioannidis, "Coronavirus Disease 2019: The Harms of Exaggerated Information and Non-Evidence-Based Measures," *European Journal of Clinical Investigation* 50, no. 4 (April 2020): e13222, https://doi.org/10.1111/eci.13222.

(28) Sucharit Bhakdi, open letter to Angela Merkel, March 26, 2020, PDF available to download until March 31, 2021, https://c.gmx.net/@824224682608695698/cI1TagSeQmioWlXK-m8vWA.

(29) Patrick Gensing and Markus Grill, "40 Prozent mehr Tests in Deutschland," *Tagesschau* (Hamburg), May 6, 2020, https://www.tagesschau.de/investigativ/corona-tests-rki-101.html.

(30) Julia Bernewasser, "Das sind die ersten Lehren der Heinsberg-Studie," *Der Tagesspeigel* (Berlin), April 9, 2020, https://www.tagesspiegel.de/wissen/zwischenergebnis-zurcoronavirus-uebertragung-das-sind-die-ersten-lehrenderheinsberg-studie/25730138.html.

(31) Paula Schneider, "'Unwissenschaftlich': Statistikerin zerlegt Heinsberg-Studie, auf die sich Laschet stützt," *Focus* (Munich), April 15, 2020, https://www.focus.de/gesundheit/news/hoffe-dass-wir-darausnur-wenig-ueber-corona-lernen-statistikerin-zerlegtheinsbergstudie-keine-transparenz-kein-wissenschaftlicher-standard_id_11881853.html.

(32) Hendrik Streeck et al., "Infection Fatality Rate of SARS-CoV-2 Infection in a German Community with a Super-Spreading Event," preprint, *medRxiv*, June 2, 2020, https://doi.org/10.1101/2020.05.04.20090076.

(33) "Field Briefing: Diamond Princess COVID-19 Cases," National Institute of Infectious Diseases (Japan), February 19, 2020, https://www.niid.go.jp/niid/en/2019-ncov-e/9407-covid-dpfe-01.html.

(34) Kenji Mizumoto et al., "Estimating the Asymptomatic Proportion of Coronavirus Disease 2019 (COVID-19) Cases on Board the Diamond Princess Cruise Ship, Yokohama, Japan, 2020," *Eurosurveillance* 25, no. 10 (March 2020): 2000180, https://doi.org/10.2807/1560-7917.ES.2020.25.10.2000180.

(35) Tara John, "Iceland Lab's Testing Suggests 50% of Coronavirus Cases Have No Symptoms," *CNN*, April 3, 2020, https://edition.cnn.com/2020/04/01/europe/iceland-testing-coronavirus-intl/index.html.

(36) Rongrong Yang, Xien Gui, and Yong Xiong, "Comparison of Clinical Characteristics of Patients with Asymptomatic vs Symptomatic Coronavirus Disease 2019 in Wuhan, China," *JAMA Network Open* 3, no. 5 (May 2020): e2010182, https://doi.org/10.1001/jamanetworkopen.2020.10182.

(37) "Erster Todesfall in Schleswig-Holstein," *Der Spiegel*, March 17, 2020, https://www.spiegel.de/wissenschaft/coronavirus-erster-todesfall-in-schleswig-holstein-a-6db5f0b0-b662-45b0-bdb4-603684d4dc92.

(38) Bettina Mittelacher, "Mediziner: Alle Corona-Toten in Hamburg waren vorerkrankt," *Berliner Morgenpost*, April 27, 2020, https://www.morgenpost.de/vermischtes/article228994571/Rechtsmediziner-Alle-Corona-Toten-hattenVorerkrankungen.html.

(39) Dominic Wichmann et al., "Autopsy Findings and Venous Thromboembolism in Patients with COVID-19: A Prospective Cohort Study," *Annals of Internal Medicine* 173, no. 4 (August 2020): 268–77, https://doi.org/10.7326/M20-2003.

(40) Nikita Jolkver, "Coronavirus: Was die Toten über COVID-19 verraten," *DW Akademie* (Bonn), April 30, 2020, https://p.dw.com/p/3baZF.

(41) SARS-CoV-2 Surveillance Group, *Characteristics of SARS-CoV-2 Patients Dying in Italy*, report based on available data on July 9, 2020, https://www.epicentro.iss.it/en/coronavirus/bollettino/Report-COVID-2019_9_july_2020.pdf.

(42) O. Haferkamp and H. Matthys, "Grippe und Lungenembolien," *Deutsche Medizinische Wochenschrift* 95, no. 51 (1970): 2560–63, https://doi.org/10.1055/s-0028-1108874.

(43) Sarah Newey, "Why Have So Many Coronavirus Patients Died in Italy?," *Telegraph*, March 23, 2020, https://www.telegraph.co.uk/global-health/science-and-disease/have-many-coronavirus-patients-died-italy.

(44) Gregory Beals, "Official Coronavirus Death Tolls Are Only an Estimate, and That Is a Problem," *NBC News*, April 15, 2020, https://www.nbcnews.com/news/world/official-coronavirus-death-tolls-are-only-estimate-problem-n1183756.

(45) Karolina Meta Beisel, "Warum Belgien die höchste Todesrate weltweit hat," *Tages-Anzeiger* (Zurich), April 22, 2020, https://

www.tagesanzeiger.ch/warum-belgien-die-hoechstetodesrate
-weltweit-hat-825753123788.

(46) John P. A. Ioannidis, Cathrine Axfors, and Despina G.
Contopoulos-Ioannidis, "Population-Level COVID-19 Mortality
Risk for Non-Elderly Individuals Overall and for Non-Elderly
Individuals without Underlying Diseases in Pandemic Epicenters,"
Environmental Research 188 (September 2020): 109890, https://
doi.org/10.1016/j.envres.2020.109890.

(47) "GrippeWeb," Robert Koch-Instituts, https://grippeweb.rki.de.

(48) "Coronavirus Disease 2019 (COVID-19): Situation Report—46,"
World Health Organization, March 6, 2020, https://www.who
.int/docs/default-source/coronaviruse/situation-reports
/20200306-sitrep-46-covid-19.pdf.

(49) "Häufig gestellte Fragen und Antworten zur Grippe," Robert
Koch-Instituts, updated January 30, 2019, https://www.rki.de
/SharedDocs/FAQ/Influenza/FAQ_Liste.html.

(50) "30 000 Tote—die kann's auch bei saisonaler Grippe geben,"
Ärzte-Zeitung (Neu-Isenburg), September 3, 2009, https://www
.aerztezeitung.de/Medizin/30000-Tote-die-kannsauch-bei-saisonaler
-Grippe-geben-371174.html.

(51) "Grippewelle war tödlichste in 30 Jahren," *Deutsches Ärzteblatt*
(Berlin), September 30, 2019, https://www.aerzteblatt.de
/nachrichten/106375/Grippewellewar-toedlichste-in-30-Jahren.

(52) "Gesundheitsministerin erklärt Grippewelle 2018 in Bayern für
beendet," *Augsburger Allgemeine*, May 10, 2018, https://www
.augsburger-allgemeine.de/wissenschaft/Gesundheitsminister
in-erklaert-Grippewelle-2018-in-Bayernfuerbeendet-id4275
0551.html.

(53) World Health Organization, "Up to 650 000 People Die of
Respiratory Diseases Linked to Seasonal Flu Each Year," news
release, December 13, 2017, https://www.who.int/mediacentre
/news/statements/2017/flu/en.

(54) Euronews, "Coronavirus in Deutschland: Sterberate steigt, RKI
erwartet zweite welle," *Euronews* (Lyon), May 5, 2020, https://
de.euronews.com/2020/05/05/coronavirus-in-deutschland
-sterberate-steigt-rki-erwartet-zweite-welle.

(55) Silvia Stringhini et al., "Repeated Seroprevalence of Anti-SARS-
CoV-2 IgG Antibodies in a Population-Based Sample from
Geneva, Switzerland," preprint, *medRxiv*, May 6, 2020, https://
doi.org/10.1101/2020.05.02.20088898.

(56) Asako Doi et al., "Estimation of Seroprevalence of Novel
Coronavirus Disease (COVID-19) Using Preserved Serum at an
Outpatient Setting in Kobe, Japan: A Cross-Sectional Study,"
preprint, *medRxiv*, May 5, 2020, https://doi.org/10.1101/2020
.04.26.20079822.

(57) Alberto L. Garcia-Basteiro et al., "Seroprevalence of Antibodies Against SARS-CoV-2 Among Health Care Workers in a Large Spanish Reference Hospital," preprint, *medRxiv*, May 2, 2020, https://doi.org/10.1101/2020.04.27.20082289.

(58) "Coronavirus: los primeros datos de seroprevalencia estiman que un 5% de la población ha estado contagiada, con variabilidad según provincias," Instituto de Salud Carlos III, Gobierno de España, May 13, 2020, https://www.isciii.es/Noticias/Noticias/Paginas/Noticias/PrimerosDatos EstudioENECOVID19.aspx.

(59) Kieran Corcoran, "A Test of 200 People Just outside Boston Found That 32% Had Been Exposed to the Coronavirus, Compared to an Official Rate of 2%," *Business Insider*, April 19, 2020, https://www.businessinsider.com/coronavirus-test-200-chelsea-massachusetts-finds-32-percent-exposed-2020-4.

(60) John Ioannidis, "The Infection Fatality Rate of COVID-19 Inferred from Seroprevalence Data," preprint, *medRxiv*, July 14, 2020, https://doi.org/10.1101/2020.05.13.20101253.

(61) Maryam Shakiba et al., "Seroprevalence of COVID-19 Virus Infection in Guilan Province, Iran," preprint, *medRxiv*, May 1, 2020, https://doi.org/10.1101/2020.04.26.20079244.

(62) Eran Bendavid et al., "COVID-19 Antibody Seroprevalence in Santa Clara County, California," preprint, *medRxiv*, posted April 30, 2020, https://doi.org/10.1101/2020.04.14.20062463.

(63) Christian Erikstrup et al., "Estimation of SARS-CoV-2 Infection Fatality Rate by Real-Time Antibody Screening of Blood Donors," *Clinical Infectious Diseases* ciaa849 (June 2020): https://doi.org/10.1093/cid/ciaa849.

(64) Fadoua Balabdaoui and Dirk Mohr, "Age-Stratified Model of the COVID-19 Epidemic to Analyze the Impact of Relaxing Lockdown Measures: Nowcasting and Forecasting for Switzerland," preprint, *medRxiv*, May 13, 2020, https://doi.org/10.1101/2020.05.08.20095059.

(65) "Preliminary Results of USC-LA County COVID-19 Study Released," University of Southern California, April 20, 2020, https://pressroom.usc.edu/preliminary-results-of-usc-la-county-covid-19-study-released.

(66) Lionel Roques et al., "Using Early Data to Estimate the Actual Infection Fatality Ratio from COVID-19 in France," *Biology* 9, no. 5 (May 2020): 97, https://doi.org/10.3390/biology9050097.

(67) Carson C. Chow et al., "Global Prediction of Unreported SARS-CoV2 Infection from Observed COVID-19 Cases," preprint, *medRxiv*, May 5, 2020, https://doi.org/10.1101/2020.04.29.20083485.

(68) Siuli Mukhopadhyay and Debraj Chakraborty, "Estimation of

Undetected COVID-19 Infections in India," preprint, *medRxiv*, May 3, 2020, https://doi.org/10.1101/2020.04.20.20072892.

(69) Robert Verity et al., "Estimates of the Severity of Coronavirus Disease 2019: A Model-Based Analysis," *Lancet: Infectious Diseases* 20, no. 6 (June 2020): 669–77, https://doi.org/10.1016/S1473-3099(20)30243-7.

(70) Kenji Mizumoto, Katsushi Kagaya, and Gerardo Chowell, "Early Epidemiological Assessment of the Transmission Potential and Virulence of Coronavirus Disease 2019 (COVID-19) in Wuhan City, China, January–February, 2020," *BMC Medicine* 18 (2020): article 217, https://doi.org/10.1186/s12916-020-01691-x.

(71) Timothy W. Russell et al., "Estimating the Infection and Case Fatality Ratio for Coronavirus Disease (COVID-19) Using Age-Adjusted Data from the Outbreak on the Diamond Princess Cruise Ship, February 2020," *Eurosurveillance* 25, no. 12 (March 2020): 2000256, https://doi.org/10.2807/1560-7917.ES.2020.25.12.2000256.

(72) "Komplikationen," Lungenärzte im Netz, https://www.lungenaerzte-im-netz.de/krankheiten/grippe/komplikationen.

(73) "103-jährige Italienerin erholt sich von COVID-19," *Donaukurier* (Ingolstadt), updated April 16, 2020, video, https://www.donaukurier.de/nachrichten/panorama/103-jaehrige-Italienerin-erholt-sich-von-COVID-19;art154670,4548023.

(74) Jason Oke and Carl Heneghan, "Global COVID-19 Case Fatality Rates," Oxford COVID-19 Evidence Service, Centre for Evidence-Based Medicine, updated June 9, 2020, https://www.cebm.net/covid-19/global-covid-19-case-fatality-rates.

(75) "113-jährige Spanierin überlebt Coronavirus-Infektion," *GMX*, May 13, 2020, https://www.gmx.net/magazine/panorama/113-jaehrigespanierin-ueberlebt-coronavirus-infektion-34698438.

(76) "SARS-CoV-2 Steckbrief zur Coronavirus-Krankheit-2019 (COVID-19)," Robert Koch-Institut, updated August 21, 2020, https://www.rki.de/DE/Content/InfAZ/N/Neuartiges_Coronavirus/Steckbrief.html.

(77) "Amtliches Dashboard COVID19," Bundesministerium für Soziales, Gesundheit, Pflege und Konsumentenschutz, https://info.gesundheitsministerium.at/dashboard_GenTod.html.

(78) "COVID-19 Daily Deaths," National Health Service (UK), https://www.england.nhs.uk/statistics/statistical-work-areas/covid-19-daily-deaths.

(79) Kathryn A. Myers and Donald R. E. Farquhar, "Improving the Accuracy of Death Certification," *CMAJ* 158, no. 10 (May 1998): 1317–23, https://www.cmaj.ca/content/cmaj/158/10/1317.full.pdf.

(80) Sandy McDowell, "Understanding Cancer Death Rates," American Cancer Society, January 25, 2019, https://www.cancer.org/latest-news/understanding-cancer-death-rates.html.

(81)   Yoon K. Loke and Carl Heneghan, "Why No-One Can Ever Recover from COVID-19 in England—A Statistical Anomaly," Centre for Evidence-Based Medicine, July 16, 2020, https:// www.cebm.net/covid-19/why-no-one-can-ever-recover-from -covid-19-in-england-a-statistical-anomaly.

(82)   "Corona—Blog," Krefeld, https://www.krefeld.de/de/inhalt /corona-aktuelle-meldungen/.

(83)   Victor G. Puelles et al., "Multiorgan and Renal Tropism of SARS-CoV-2," *New England Journal of Medicine* 383 (August 2020): 590–92, https://doi.org/10.1056/NEJMc2011400.

(84)   David Bainton, Glynne R. Jones, and David Hole, "Influenza and Ischaemic Heart Disease—A Possible Trigger for Acute Myocardial Infarction?," *International Journal of Epidemiology* 7, no. 3 (September 1978): 231–39, https://doi.org/10.1093/ije/7.3.231.

(85)   Hiroshi Kido et al., "Role of Host Cellular Proteases in the Pathogenesis of Influenza and Influenza-Induced Multiple Organ Failure," *Biochimica et Biophysica Acta (BBA) – Proteins and Proteomics* 1824, no. 1 (January 2012): 186–94, https://doi .org/10.1016/j.bbapap.2011.07.001.

(86)   Debby van Riel, Rob Verdijk, and Thijs Kuiken, "The Olfactory Nerve: A Shortcut for Influenza and Other Viral Diseases into the Central Nervous System," *Journal of Pathology* 235, no. 2 (January 2015): 277–87, https://doi.org/10.1002/path.4461.

(87)   Camilla Rothe et al., "Transmission of 2019-nCoV Infection from an Asymptomatic Contact in Germany," *New England Journal of Medicine* 382 (March 2020): 970–71, https://doi .org/10.1056/NEJMc2001468.

(88)   Kai Kupferschmidt, "Study Claiming New Coronavirus Can Be Transmitted by People without Symptoms Was Flawed," *Science*, February 3, 2020, https://www.sciencemag.org/news/2020/02 /paper-non-symptomatic-patient-transmitting-coronavirus-wrong.

(89)   Russell M. Viner et al., "School Closure and Management Practices During Coronavirus Outbreaks Including COVID-19: A Rapid Systematic Review," *Lancet* 4, no. 5 (May 2020): 397–404, https://doi.org/10.1016/S2352-4642(20)30095-X.

(90)   Pamela Dörhöfer, "Italien leidet unter dem Coronavirus: Sterberate ist erschreckend hoch," *Frankfurter Rundschau*, April 14, 2020, https://www.fr.de/panorama/coronavirus-SARS-CoV -2-sterberate-italien-deutlich-hoeher-rest-welt-zr-13604897.html.

(91)   Ernesto Diffidenti, "Coronavirus, i contagiati reali in Italia sono almeno 100mila," *Il Sole 24 Ore* (Milan), March 17, 2020, https://www.ilsole24ore.com/art/coronavirus-contagiati-realiin -italia-sono-almeno-100mila-ADnzowD.

(92)   Kat Lay, "Coronavirus: Record Weekly Death Toll as Fearful Patients Avoid Hospitals," *Times* (UK), April 15, 2020, https://

www.thetimes.co.uk/article/coronavirus-record-weeklydeath
-toll-as-fearful-patients-avoid-hospitals-bm73s2tw3.

(93) Paul Nuki, "Two New Waves of Deaths Are about to Break over
the NHS, New Analysis Warns," *Telegraph*, April 25, 2020,
https://www.telegraph.co.uk/global-health/science-and-disease
/two-new-waves-deaths-break-nhs-new-analysis-warns.

(94) Ceylan Yeinsu, "N.H.S. Overwhelmed in Britain, Leaving
Patients to Wait," *New York Times*, January 3, 2018, https://
www.nytimes.com/2018/01/03/world/europe/uk-national-health
-service.html.

(95) Denis Campbell, "Health Services Overloaded Despite Support
Pledges, Claims Report," *Guardian* (US edition), May 20, 2018,
https://www.theguardian.com/politics/2018/may/21/health
-services-overloaded-despite-support-pledges-claims-report.

(96) Michael Savage, "NHS Winter Crisis Fears Grow after
Thousands of EU Staff Quit," *Guardian* (US edition), November
24, 2019, https://www.theguardian.com/society/2019/nov/24
/nhs-winter-crisis-thousands-eu-staff-quit.

(97) Amanda MacMillan, "Hospitals Overwhelmed by Flu Patients
Are Treating Them in Tents," *Time*, January 18, 2018, https://
time.com/5107984/hospitals-handling-burden-flupatients.

(98) Helen Branswell, "A Severe Flu Season Is Stretching Hospitals
Thin. That Is a Very Bad Omen," *STAT*, January 15, 2018,
https://www.statnews.com/2018/01/15/flu-hospital-pandemics.

(99) "Coronavirus Fact-Check #1: 'COVID19 Is Having an
Unprecedented Impact on ICUs," *OffGuardian*, April 2, 2020,
https://off-guardian.org/2020/04/02/coronavirus-fact-check-1
-flu-doesnt-overwhelm-our-hospitals.

(100) R. Salamanca, "La gripe colapsa los hospitales de media España,"
El Mundo (Madrid), January 12, 2017, https://www.elmundo.es
/ciencia/2017/01/12/58767cb4268e3e1f448b459a.html.

(101) Daniel Ventura, "¿Por qué la gripe significa colapso en los
hospitales españoles?," *HuffPost* (Spain edition), January 13,
2017, https://www.huffingtonpost.es/2017/01/13/gripe-colapso
hospitales_n_14135402.html.

(102) Simona Ravizza, "Milano, terapie intensive al collasso per
l'influenza: già 48 malati gravi molte operazioni rinviate,"
Corriere della Sera (Milan), January 10, 2018, https://milano
.corriere.it/notizie/cronaca/18_gennaio_10/milano-terapie
-intensive-collasso-l-influenza-gia-48-malati-gravi-molte
-operazioni-rinviate-c9dc43a6-f5d1-11e7-9b06-fe054c3be
5b2.shtml.

(103) Christian Baars, "Mehr Tote durch resistente Keime," *Tagesschau*
(Hamburg), November 18, 2019, https://www.tagesschau.de
/inland/antibiotika-keime-resistent-101.html.

(104) "Europäische Union: Altersstruktur in den Mitgliedsstaaten im Jahr 2019," Statista, March 2020, https://de.statista.com /statistik/daten/studie/248981/umfrage/altersstruktur-in-den-eu -laendern.

(105) Savannah Blank, "Wieso sterben in Italien so viele an COVID-19 und wieso sind so viele infiziert?," *Südwest Presse* (Ulm), April 30, 2020, https://www.swp.de/panorama/coronavirus-italien -aktuellwieso-sterben-in-italien-so-viele-an-corona-wieso-hat -italiensoviele-infizierte-zahlen-tote-gruende-45080326.html.

(106) Stefania Boccia, Walter Ricciardi, and John P. A. Ioannidis, "What Other Countries Can Learn from Italy During the COVID-19 Pandemic," *JAMA Internal Medicine* 180, no. 7 (July 2020): 927–28, https://doi.org/10.1001/jamainternmed .2020.1447.

(107) Paul Kreiner, "Beim Smog ist Italien das China Europas," *Der Tagesspiegel* (Berlin), December 2, 2015, https://www. tagesspiegel.de/gesellschaft/panorama/luftverschmutzung-beim -smog-ist-italien-daschinaeuropas/12668866.html.

(108) Marco Martuzzi et al., *Health Impact of PM10 and Ozone in 13 Italian Cities* (Copenhagen: World Health Organization, 2006), http://www.euro.who.int/__data/assets/pdf_file/0012/91110 /E8870o.pdf.

(109) Daniel P. Croft et al., "The Association between Respiratory Infection and Air Pollution in the Setting of Air Quality Policy and Economic Change," *Annals of the American Thoracic Society* 16, no. 3 (March 2019): https://doi.org/10.1513/Annals ATS.201810-691OC.

(110) Xiao Wu et al., "Exposure to Air Pollution and COVID-19 Mortality in the United States: A Nationwide Cross-Sectional Study," preprint, *medRxiv*, April 27, 2020, https://doi.org /10.1101/2020.04.05.20054502.

(111) Susanne Pfaller and Matthias Lauer, "Trauer in Corona-Zeiten: Mehr Anzeigen und Feuerbestattungen," *Bayerischer Rundfunk*, April 29, 2020, https://www.br.de/nachrichten/bayern/trauer-in -corona-zeitenmehr-anzeigen-und-feuerbestattungen,RxZCWs0.

(112) Katja Thorwarth, "New Yorker Notarzt über Corona-Krise in der Bronx: „Manchmal 200 Erkrankungen in einem Stockwerk,"" *Frankfurter Rundschau*, May 14, 2020, https:// www.fr.de/politik/coronavirus-corona-krise-usanotarzt-lage -new-york-bronx-zr-13762623.html.

(113) Lucia De Franceschi et al., "Acute Hemolysis by Hydroxycloroquine Was Observed in G6PD-Deficient Patient with Severe COVID-19 Related Lung Injury," *European Journal of Internal Medicine* 77 (July 2020): 136–37, https://doi.org /10.1016/j.ejim.2020.04.020.

(114) "Glucose-6-Phosphate Dehydrogenase Deficiency," Genetics Home
       Reference, National Institutes of Health, August 17, 2020, https://
       ghr.nlm.nih.gov/condition/glucose-6-phosphate-dehydrogenase
       -deficiency#statistics.

(115) "Charité-Chefvirologe warnt vor dramatischer Corona-Welle im
       Herbst," *BZ* (Berlin), updated April 1, 2020, https://www
       .bz-berlin.de/berlin/charite-chefvirologe-warnt-vor-dramatischer
       -corona-welle-im-herbst.

(116) Dominik Straub, "Letalität in Deutschland 30-mal niedriger
       als in Italien – wie ist das möglich?," *Der Tagesspiegel* (Berlin),
       March 11, 2020, https://www.tagesspiegel.de/politik/coronavirus
       -in-europaletalitaet-in-deutschland-30-mal-niedriger-als-in
       -italien-wieistdas-moeglich/25626678.html.

(117) BMG, Twitter post, March 14, 2020, 6:55 a.m., https://twitter
       .com/bmg_bund/status/1238780849652465664.

(118) "Coronavirus Restrictions: What's Closed (and What's Open) in
       Germany?," *Local* (German edition), updated March 20, 2020,
       https://www.thelocal.de/20200316/coronavirus-restrictions
       -whats-closed-and-whats-open-in-germany.

(119) John P. A. Ioannidis, "A Fiasco in the Making? As the Coronavirus
       Pandemic Takes Hold, We Are Making Decisions without Reliable
       Data," *STAT*, March 17, 2020, https://www.statnews.com
       /2020/03/17/a-fiasco-in-the-making-as-the-coronavirus-pandemic
       -takes-hold-we-are-making-decisions-without-reliable-data.

(120) Christian Baars, "Radikale Maßnahmen für viele Monate?,"
       Tagesschau (Hamburg), March 17, 2020, https://www.tagesschau
       .de/investigativ/ndr/coronavirus-studie-london-101.html.

(121) Matt Ridley and David Davis, "Is the Chilling Truth That
       the Decision to Impose Lockdown Was Based on Crude
       Mathematical Guesswork?," *Telegraph*, May 10, 2020, https://
       www.telegraph.co.uk/news/2020/05/10/chilling-truth-decision
       -impose-lockdown-based-crude-mathematical.

(122) "„Exponentielles Wachstum": RKI mahnt eindringlich zum
       Abstandhalten," *Deutsches Ärzteblatt* (Berlin), March 20, 2020,
       https://www.aerzteblatt.de/nachrichten/111209/Exponentielles
       -Wachstum-RKI-mahnt-eindringlich-zum-Abstandhalten.

(123) Matthias an der Heiden and Udo Buchholz, *Modellierung
       von Beispielszenarien der SARS-CoV-2-Epidemie 2020 in
       Deutschland*, Robert Koch Institut, March 20, 2020, https://doi
       .org/10.25646/6571.2.

(124) "Deutsche Krankenhäuser nehmen COVID-19-Patienten aus
       Italien und Frankreich auf," *Deutsches Ärzteblatt* (Berlin),
       March 24, 2020, https://www.aerzteblatt.de/nachrichten
       /111286/Deutsche-Krankenhaeuser-nehmen-COVID-19
       -Patienten-aus-Italien-undFrankreich-auf.

(125) „„Verdopplungszeit" zehn Tage—So weit ist Deutschland von Merkels Ziel entfernt," *Welt* (Berlin), March 3, 2020, https://www.welt.de/politik/deutschland/article206895285/Coronavirus-So-weit-ist-Deutschland-von-Merkels-Zielvorgabeentfernt.html.

(126) Ulrich Stoll and Christian Rohde, "Zwischen 'schneller Kontrolle' und 'Anarchie,'" *ZDF Heute* (Mainz), March 31, 2020, https://www.zdf.de/nachrichten/politik/f21-corona-dokument-innenministerium-100.html.

(127) Thomas Steinmann, "Innenministerium warnt vor Wirtschaftscrash," *Capital* (Hamburg), April 1, 2020, https://www.capital.de/wirtschaft-politik/innenministerium-warnt-vor-wirtschaftscrash.

(128) Matthias an der Heiden and Osamah Hamouda, "Schätzung der aktuellen Entwicklung der SARS-CoV-2-Epidemie in Deutschland—Nowcasting," *Epidemiologisches Bulletin* 17, (April 2020): 10–16, https://doi.org/10.25646/6692.4.

(129) Christian Geinitz, "„In den Kliniken stehen Tausende Betten leer"," *Frankfurter Allgemeine*, updated April 15, 2020, https://www.faz.net/aktuell/wirtschaft/fehlplanung-der-politikin-den-kliniken-stehen-betten-leer-16725981.html.

(130) *Using Face Masks in the Community* (Stockholm: European Centre for Disease Prevention and Control, 2020), https://www.ecdc.europa.eu/sites/default/files/documents/COVID-19-use-face-masks-community.pdf.

(131) Samy Rengasamy, Benjamin Eimer, and Ronald E. Shaffer, "Simple Respiratory Protection—Evaluation of the Filtration Performance of Cloth Masks and Common Fabric Materials Against 20–1000 nm Size Particles," *Annals of Occupational Hygiene* 54, no. 7 (October 2010): 789–98, https://doi.org/10.1093/annhyg/meq044.

(132) C. Raina MacIntyre et al., "A Cluster Randomised Trial of Cloth Masks Compared with Medical Masks in Healthcare Workers," *BMJ Open* 5, no. 4 (2015): e006577, https://doi.org/10.1136/bmjopen-2014-006577.

(133) "Advice on the Use of Masks in the Context of COVID-19," World Health Organization, June 5, 2020, https://www.who.int/publications/i/item/advice-on-the-use-of-masks-in-the-community-during-home-care-and-in-healthcare-settings-in-the-context-of-the-novel-coronavirus-(2019-ncov)-outbreak.

(134) Timo Mitze et al., *Face Masks Considerably Reduce COVID-19 Cases in Germany: A Synthetic Control Method Approach* (Bonn, Germ.: Institute of Labor Economics, 2020), http://ftp.iza.org/dp13319.pdf.

(135) Udo Buchholz, Silke Buda, and Kerstin Prahm, "Abrupter Rückgang der Raten an Atemwegserkrankungen in der deutschen

Bevölkerung," *Epidemiologisches Bulletin* 16 (April 2020): 7–9, https://doi.org/10.25646/6636.2.

(136) Denis G. Rancourt, "Masks Don't Work: A Review of Science Relevant to COVID-19 Social Policy," *River Cities' Reader* (IA), June 11, 2020, https://www.rcreader.com/commentary/masks-dont -work-covid-a-review-of-science-relevant-to-covide-19-social-policy.

(137) Pietro Vernazza, "Atemschutzmasken für alle—Medienhype oder unverzichtbar?," Klinik für Infektiologie/Spitalhygiene, Kantonsspital St. Gallen, April 5, 2020, https://infekt.ch/2020/04 /atemschutzmasken-fuer-alle-medienhype-oder-unverzichtbar.

(138) Klaus Wedekind, "Drosten warnt vor zweiter Corona-Welle," *n-tv* (Cologne), April 20, 2020, https://www.n-tv.de/panorama /Drosten-warnt-vor-zweiter-Corona-Welle-article21726926.html.

(139) Joachim Czichos, "Erst Bakterien führten zur tödlichen Katastrophe," *Welt* (Berlin), August 11, 2008, https://www.welt .de/gesundheit/article2295849/Erst-Bakterienfuehrten-zur -toedlichen-Katastrophe.html.

(140) Marie E. Killerby et al., "Human Coronavirus Circulation in the United States 2014–2017," *Journal of Clinical Virology* 101 (April 2018): 52–56, https://doi.org/10.1016/j.jcv.2018.01.019.

(141) Mika J. Mäkelä et al., "Viruses and Bacteria in the Etiology of the Common Cold," *Journal of Clinical Microbiology* 36, no. 2 (February 1998): 539–42, https://doi.org/10.1128/JCM.36.2 .539-542.1998.

(142) "Was Sie über die Grafik wissen sollten, über die Deutschland spricht," *Welt* (Berlin), April 26, 2020, https://www.welt.de /wissenschaft/article207456203/Coronavirus-Stefan-Homburg -und-die-Grafik-ueber-die-Deutschlandspricht.html.

(143) "Virologe Drosten warnt: Deutsche könnten Corona-Vorsprung verspielen," *Stern*, video, 1:04, April 23, 2020, https://www .stern.de/gesundheit/news-im-video--drosten-warnt---deutsche -koennten-corona-vorsprungverspielen9236028.html.

(144) "RKI warnt in Coronavirus-Krise: Reproduktionszahl wieder über kritischem Wert 1," *MSN* (Germany), May 10, 2020, https://www.msn.com/de-de/nachrichten/coronavirus/rkiwarnt -in-coronavirus-krise-reproduktionszahl-wiederüberkritischem -wert-1/ar-BB13RlEi.

(145) "Sterbefallzahlen in Deutschland steigen an," *Süddeutsche Zeitung*, April 30, 2020, https://www.sueddeutsche.de/gesund heit/coronavirenuebersterblichkeit-COVID-19-statistischesbunde samt-1.4893709.

(146) Jürgen Mladek, "Seehofer stellt Corona-Kritiker kalt," *Nordkurier* (Neubrandenburg), May 14, 2020, https://www .nordkurier.de/politik-und-wirtschaft/seehofer-stellt-corona -kritiker-kalt-1439370305.html.

(147) Seth Flaxman et al., "Estimating the Effects of Non-Pharmaceutical Interventions on COVID-19 in Europe," *Nature* 584 (2020): 257–61, https://doi.org/10.1038/s41586-020-2405-7.

(148) Comments section below Flaxman et al., "Estimating the Effects of Non-Pharmaceutical Interventions," https://www.nature.com/articles/s41586-020-2405-7#article-comments.

(149) "Why Is Denmark Not Recommending Face Masks to the Public?," *Local* (Denmark), May 11, 2020, https://www.thelocal.dk/20200511/why-is-denmark-not-recommending-face-masks-to-the-public.

(150) Hildburg Bruns, "Berlins erste Corona-Klinik," *Bild* (Berlin), May 11, 2020, https://www.bild.de/regional/berlin/berlin-aktuell/corona-klinik-in-berlin-fertig-knapp-500-betten-im-stand-by-modus-70577074.bild.html.

(151) Uwe Janssens, ",,Wir haben genug Intensivbetten"," interview by Arndt Reuning, *Deutschlandfunk* (Cologne), March 11, 2020, https://www.deutschlandfunk.de/corona-notfallplaene-in-krankenhaeusern-wir-habengenug.676.de.html?dram:article_id=472287.

(152) "Überlastung deutscher Krankenhäuser durch COVID-19 laut Experten unwahrscheinlich," *Deutsches Ärzteblatt* (Berlin), March 12, 2020, https://www.aerzteblatt.de/nachrichten/111029/Ueberlastungdeutscher-Krankenhaeuser-durch-COVID-19-lautExpertenunwahrscheinlich.

(153) Barbara Gillmann, "RKI: Zahl der Intensivbetten wird nicht reichen," *Handelsblatt* (Düsseldorf), April 3, 2020, https://www.handelsblatt.com/politik/deutschland/corona-epidemie-rki-zahl-der-intensivbetten-wirdnichtreichen/25712008.html?ticket=ST-3691123-xCgN9jb0yWPZsyeB97s7-ap5.

(154) Bundesministerium des Innern, für Bau und Heimat, *Wie wir COVID-19 unter Kontrolle bekommen*, March 2020, https://www.bmi.bund.de/SharedDocs/downloads/DE/veroeffentlichungen/2020/corona/szenarienpapier-covid-19.pdf;jsessionid=8FAD89A1832ABFC4DB485C5625C8DE71.2_cid295?__blob=publicationFile&v=4.

(155) Elena Kuch, Jennifer Lange, and Christoph Prössl, "Kurzarbeit trotz Rettungsschirm," *Tagesschau* (Hamburg), April 22, 2020, https://www.tagesschau.de/investigativ/ndr/krankenhaeuser-kurzarbeit-101.html.

(156) Kim Norvell and Jayne O'Donnell, "Thousands of US Medical Workers Furloughed, Laid Off as Routine Patient Visits Drop During Coronavirus Pandemic," *USA Today*, updated April 2, 2020, https://eu.usatoday.com/story/news/health/2020/04/02/coronavirus-pandemic-jobs-us-health-care-workersfurloughed laid-off/5102320002.

(157)   Kit Knightly, "COVID19: Are Ventilators Killing People?,"
        OffGuardian, May 6, 2020, https://off-guardian.org/2020
        /05/06/covid19-are-ventilators-killing-people.
(158)   "COVID-19: Beatmung—und dann?," DocCheck, March 31,
        2020, https://www.doccheck.com/de/detail/articles/26271
        -COVID-19-beatmung-und-dann.
(159)   Martin Gould, "EXCLUSIVE: 'It's a Horror Movie.' Nurse
        Working on Coronavirus Frontline in New York Claims the
        City Is 'Murdering' COVID-19 Patients by Putting Them on
        Ventilators and Causing Trauma to the Lungs," *Daily Mail*,
        updated May 14, 2020, https://www.dailymail.co.uk/news
        /article-8262351/Nurse-New-York-claims-city-killing-COVID-19
        -patientsputtingventilators.html.
(160)   Jochen Taßler and Jan Schmitt, "Mehr Schaden als Nutzen?,"
        Tagesschau (Hamburg), April 30, 2020, https://www.tagesschau
        .de/investigativ/monitor/beatmung-101.html.
(161)   "„Es wird zu häufig intubiert und invasiv beatmet"," *Frankfurter
        Allgemeine Zeitung*, April 7, 2020, https://www.vpneumo.de
        /fileadmin/pdf/f2004071.007_Voshaar.pdf.
(162)   Kristin Kielon, "So Funktioniert Künstliche Beatmung,"
        Mitteldeutscher Rundfunk (Leipzig), March 24, 2020, https://
        www.mdr.de/wissen/so-funktioniert-beatmung-intensivstation
        -corona-100.html.
(163)   "Modes of Transmission of Virus Causing COVID-19:
        Implications for IPC Precaution Recommendations," World
        Health Organization, March 29, 2020, https://www.who.int
        /news-room/commentaries/detail/modes-of-transmission
        -of-virus-causing-covid-19-implications-for-ipc-precaution
        -recommendations.
(164)   Neeltje van Doremalen et al., "Aerosol and Surface Stability of
        SARS-CoV-2 as Compared with SARS-CoV-1," *New England
        Journal of Medicine* 382 (April 2020): 1564–67, https://doi
        .org/10.1056/NEJMc2004973.
(165)   Young-Il Kim et al., "Infection and Rapid Transmission of SARS-
        CoV-2 in Ferrets," *Cell Host & Microbe* 27, no. 5 (May 2020):
        704–9.e2, https://doi.org/10.1016/j.chom.2020.03.023.
(166)   Matthias Thöns, "„Sehr falsche Prioritäten gesetzt und alle
        ethischen Prinzipien verletzt"," interview by Peter Sawicki,
        Deustchlandfunk, April 11, 2020, https://www.deutschlandfunk
        .de/palliativmedizinerzu-COVID-19-behandlungen-sehrfalsche
        .694.de.html?dram:article_id=474488.
(167)   David L. Katz, "Is Our Fight against Coronavirus Worse Than
        the Disease?," *New York Times*, March 20, 2020, https://www
        .nytimes.com/2020/03/20/opinion/coronaviruspandemic-social
        -distancing.html.

(168) CNN, "'We are creating a catastrophic health care situation,'" Facebook post, video, 6:09, May 2, 2020, https://www.facebook .com/cnn/posts/10160799274796509.

(169) Scott W. Atlas, "The Data Is In—Stop the Panic and End the Total Isolation," *The Hill* (Washington, DC), April 22, 2020, https://thehill.com/opinion/healthcare/494034-the-data-are-in -stop-the-panic-and-end-the-total-isolation.

(170) Robert Birnbaum and Georg Ismar, "Schäuble will dem Schutz des Lebens nicht alles unterordnen," *Der Tagesspiegel* (Berlin), April 26, 2020, https://www.tagesspiegel.de/politik /bundestagspraesident-zurcorona-krise-schaeuble-will-dem -schutz-des-lebens-nichtallesunterordnen/25770466.html.

(171) Miriam Kruse, "Menschenleben versus Menschenwürde?," *SWR Aktuell*, April 27, 2020, https://www.swr.de/swraktuell /schaueble-wertediskussion-zu-corona-100.html.

(172) Andy Sumner, Chris Hoy, and Eduardo Ortiz-Juarez, "Estimates of the Impact of COVID-19 on Global Poverty," (working paper, United Nations University World Institute for Development Economics Research, 2020), https://www.wider.unu.edu /publication/estimates-impact-covid-19-global-poverty.

(173) "Amerikas Wirtschaftsleistung sinkt um bis zu 30 Prozent," *Frankfurter Allgemeine*, updated May 18, 2020, https://www .faz.net/aktuell/wirtschaft/usa-notenbank-federwartet -dramatischen-einbruch-der-wirtschaft-16774864.html.

(174) Ines Zöttl, "US-Arbeitsmarkt in der Coronakrise: US-Arbeitsmarkt in der Coronakrise," *Spiegel* (Hamburg), May 9, 2020, https:// www.spiegel.de/wirtschaft/corona-krise-in-den-usa-der-auftakt -der-tragoedie-a-532f7a6b-3a0d-4a8f-a38d-db91ead7990b.

(175) "EU vor Rezession von 'historischem Ausmaß," *Tagesschau* (Hamburg), May 6, 2020, https://www.tagesschau.de/wirtschaft /corona-eurozone-rezession-101.html.

(176) Benjamin Bidder, "'Das wird ein Zangenangriff auf Deutschlands Wholstand,'" *Spiegel* (Hamburg), May 17, 2020, https://www .spiegel.de/wirtschaft/corona-krise-das-wird-ein-zangenangriff -auf-deutschlands-wohlstand-a-eaf27caa-342d-4aca-bcb1-e84b 15ca5a2d.

(177) Britta Beeger, "Warum die Arbeitslosigkeit steigt," *Frankfurter Allgemeine*, May 4, 2020, https://www.faz.net/aktuell/wirtschaft /corona-krise-warum-die-arbeitslosigkeit-in-deutschland-steigt -16753941.html.

(178) "Kampf gegen Corona: Größtes Hilfspaket in der Geschichte Deutschlands," Bundesministerium der Finanzen, May 22, 2020, https://www.bundesfinanzministerium.de/Content/DE/Standard artikel/Themen/Schlaglichter/Corona-Schutzschild/2020 -03-13-Milliarden-Schutzschild-fuer-Deutschland.html.

(179) Mallory Simon, "75,000 Americans at Risk of Dying from Overdose or Suicide Due to Coronavirus Despair, Group Warns," *CTV News* (Ottawa), May 8, 2020, https://www.ctvnews.ca/health/coronavirus/75-000-americans-at-risk-of-dying-from-overdose-or-suicide-due-to-coronavirus-despair-group-warns-1.4930801.

(180) Agence France-Presse, "Australia Fears Suicide Spike Due to Virus Shutdown," *Telegraph*, May 7, 2020, https://www.telegraph.co.uk/news/2020/05/07/australia-fears-suicide-spike-due-virus-shutdown.

(181) "Mehr Tote durch Schlaganfälle, Infarkte und Suizide erwartet," *BZ* (Berlin), May 7, 2020, https://www.bz-berlin.de/ratgeber/coronavirus-lockdown-mehr-tote-durch-schlaganfaelle-infarkte-und-suizide-erwartet.

(182) Catharine Paddock, "Heart Attack Risk Higher with Job Loss," *Medical News Today*, November 20, 2012, https://www.medicalnewstoday.com/articles/252985.

(183) Hinnerk Feldwisch-Drentrup, "Warum in der Coronakrise nicht nur das Virus die Gesundheit gefährdet," *Der Tagesspiegel* (Berlin), May 16, 2020, https://www.tagesspiegel.de/wissen/die-gesundheitlichenfolgen-des-lockdowns-jetzt-sind-es-30-prozentwenigerherzinfarkte-doch-spaeter-werden-es-wohl-mehr/25834148.html.

(184) COVIDSurg Collaborative, "Elective Surgery Cancellations Due to the COVID-19 Pandemic: Global Predictive Modelling to Inform Surgical Recovery Plans," *British Journal of Surgery*, May 12, 2020, https://doi.org/10.1002/bjs.11746.

(185) "Anzahl der Sterbefälle in Deutschland nach Altersgruppe im Jahr 2018," Statista, May 2020, https://de.statista.com/statistik/daten/studie/1013307/umfrage/sterbefaelle-in-deutschland-nach-alter.

(186) MarieLuise Dr. Stiefel et al., "Corona: Schützen Sie uns Ältere nicht um diesen Preis! Selbstbestimmt altern und sterben!" Change.org, PBC, https://www.change.org/p/bundeskanzlerin-corona-sch%C3%BCtzen-sie-%C3%A4ltere-nicht-um-diesen-preis-selbstbestimmt-altern-und-sterben.

(187) "UNICEF: Höhere risiken für Kinder wegen Massnahmen zur eindämmung des Coronavirus," UNICEF, March 23, 2020, https://www.unicef.de/informieren/aktuelles/presse/2020/risiken-fuer-kinder-bei-eindaemmung-des-coronavirus/213060.

(188) Peter Dabrock, ""Kinder brauchen andere Kinder"," interview by Anke Schaefer, *Deutschlandfunk Kultur* (Berlin), April 14, 2020, https://www.deutschlandfunkkultur.de/sozialethiker-kritisiertlange-kitaschliessungenkinder.1008.de.html?dram:article_id=474595.

(189) "Deutschlands Lehrer-Chef: 'Ein Viertel aller Schüler abgehängt," *Focus*, May 8, 2020, https://www.focus.de/familie /eltern/meidinger-zuschulschliessungen-deutschlands-lehrer-chef -ein-viertel-allerschuelerabgehaengt_id_11878788.html.

(190) Sue Odenthal and Martina Morawietz, "'Wir sind extrem blind im Kinderschutz,'" *ZDF Heute* (Mainz), April 28, 2020, https:// www.zdf.de/nachrichten/panorama/coronavirus-kinderschutz -jugendamt-100.html.

(191) Christoph Hein, "Auf Corona folgt der Hunger," *Frankfurter Allgemeine*, April 22, 2020, https://www.faz.net/aktuell /wirtschaft/un-warnt-auf-coronafolgt-die-hungersnot-16736443 .html.

(192) Leslie Roberts, "Why Measles Deaths Are Surging—and Coronavirus Could Make It Worse," *Nature*, updated April 9, 2020, https://www.nature.com/articles/d41586-020-01011-6.

(193) Stefan Homburg, "Warum Deutschlands Lockdown falsch ist— und Schweden vieles besser macht," *Welt* (Berlin), April 15, 2020, https://www.welt.de/wirtschaft/plus207258427 /Schwedenals-Vorbild-Finanzwissenschaftler-gegen-Corona Lockdown.html?ticket=ST-A-1309422-NghISRcCkH3oTuFUao V5-ssosignin-server.

(194) Johan Giesecke, "„Lockdown verschiebt Tote in die Zukunft"," interview by Johannes Perterer, Addendum, Quo Vadis Veritas, April 24, 2020, https://www.addendum.org/coronavirus /interview-johan-giesecke.

(195) André Anwar, "WHO lobt Sonderweg: Können wir vom Modell Schweden lernen?," *Augsburger Allgemeine*, May 5, 2020, https://www.augsburger-allgemeine.de/panorama/WHO -lobt-Sonderweg-Koennen-wir-vom-Modell-Schweden lernenid57329376.html.

(196) "Ende der Pandemie? Neue Zahlen widersprechen Regierungs- Linie- Punkt.PRERADOVIC mit Prof. Homburg," YouTube video, 16:17, interview by Punkt Preradovic, May 4, 2020, posted by "Punkt.PRERADOVIC," https://www.youtube.com /watch?v=WFkMIlKyHoI.

(197) "Wo die Coronavirus-Pandemie ohne Lockdown bewältigt wird," *Der Tagesspiegel* (Berlin), April 18, 2020, https://www .tagesspiegel.de/wissen/von-hongkong-lernenwo-die-coronavirus -pandemie-ohne-lockdownbewaeltigtwird/25752346.html.

(198) Gearoid Reidy, "Japan Was Expecting a Coronavirus Explosion. Where Is It?," *Japan Times*, March 20, 2020, https://www .japantimes.co.jp/news/2020/03/20/national/coronavirus -explosion-expected-japan/.

(199) Aria Bendix, "South Korea Has Tested 140,000 People for the Coronavirus. That Could Explain Why Its Death Rate Is Just

0.6% — Far Lower Than in China or the US," *Business Insider*, March 5, 2020, https://www.businessinsider.com/south-korea -coronavirus-testing-death-rate-2020-3.

(200) *Non-Pharmaceutical Public Health Measures for Mitigating the Risk and Impact of Epidemic and Pandemic Influenza* (World Health Organization, 2019), https://www.who.int/influenza /publications/public_health_measures/publication/en.

(201) "Nobel Prize Winning Scientist Prof Michael Levitt: Lockdown Is a 'Huge Mistake,'" YouTube video, 34:33, interview by UnHerd, posted by "UnHerd," May 2, 2020, https://www.youtube.com /watch?v=bl-sZdfLcEk.

(202) Adelina Comas-Herrera et al., *Mortality Associated with COVID-19 Outbreaks in Care Homes: Early International Evidence*, LTC Covid, updated June 26, 2020, https://ltccovid .org/2020/04/12/mortality-associated-with-covid-19-outbreaks -in-care-homes-early-international-evidence.

(203) avstoesser, "Falsche Prioritäten gesetzt und ethische Prinzipien verletzt," Pflegeethik Initiative, Deutschland e.V., April 15, 2020, http://pflegeethik-initiative.de/2020/04/15/corona-krise-falsche -prioritaeten-gesetzt-und-ethische-prinzipien-verletzt.

(204) "'No Second Wave' Despite Record Surge in New COVID-19 Cases, Says Czech Minister," *Kafkadesk*, June 29, 2020, https:// kafkadesk.org/2020/06/29/no-second-wave-despite-record -surge-in-new-covid-19-cases-says-czech-minister.

(205) "Mandatory Face Masks Should Not Pose Problems, Says Swiss Train Boss," *Swissinfo* (Bern), July 3, 2020, https://www.swiss info.ch/eng/mandatory-face-masks-should-not-pose-problems --says-swiss-train-boss/45879254.

(206) Daland Segler, "„Anne Will": Wie hart trifft uns die „neue Normalität"?," *Frankfurter Rundschau*, May 29, 2020, https:// www.fr.de/kultur/tv-kino/corona-talk-anne-will-ardhart-trifft -neue-normalitaet-zr-13667631.html.

(207) "Corona-Folgen bekämpfen, Wohlstand sichern, Zukunftsfähigkeit stärken," Federal Ministry of Finance (Germany), June 3, 2020, https://www.bundesfinanzministerium .de/Content/DE/Standardartikel/Themen/Schlaglichter /Konjunkturpaket/2020-06-03-eckpunktepapier.pdf?__blob =publicationFile&v=10.

(208) "'Es braucht eine globale Anstrengung,'" *Tagesschau* (Hamburg), April 12, 2020, https://www.tagesschau.de/ausland/gates-corona -101.html.

(209) Jincun Zhao et al., "Airway Memory CD4+ T Cells Mediate Protective Immunity against Emerging Respiratory Coronaviruses," *Immunity* 44, no. 6 (June 2016): 1379–91, https://doi.org/10.1016/j.immuni.2016.05.006.

(210) Annika Nelde et al., "SARS-CoV-2 T-cell Epitopes Define Heterologous and COVID-19-Induced T-Cell Recognition," preprint, posted June 17, 2020, https://doi.org/10.21203/rs .3.rs-35331/v1.

(211) Alba Grifoni et al., "Targets of T-Cell Responses to SARS-CoV-2 Coronavirus in Humans with COVID-19 Disease and Unexposed Individuals," *Cell* 181, no. 7 (June 2020): 1489–501.e15, https:// doi.org/10.1016/j.cell.2020.05.015.

(212) Takuya Sekine et al., "Robust T Cell Immunity in Convalescent Individuals with Asymptomatic or Mild COVID-19," *Cell*, (August 2020), https://doi.org/10.1016/j.cell.2020.08.017.

(213) Barry Rockx et al., "Comparative Pathogenesis of COVID-19, MERS, and SARS in a Nonhuman Primate Model," *Science* 368, no. 6494 (May 2020): 1012–15, https://doi.org/10.1126/science .abb7314.

(214) Shibo Jiang, "Don't Rush to Deploy COVID-19 Vaccines and Drugs without Sufficient Safety Guarantees," *Nature* 579 (2020): 321, https://doi.org/10.1038/d41586-020-00751-9.

(215) Sanjay Kumar, "Scientists Scoff at Indian Agency's Plan to Have COVID-19 Vaccine Ready for Use Next Month," *Science*, July 6, 2020, https://www.sciencemag.org/news/2020/07/scientists -scoff-indian-agencys-plan-have-COVID-19-vaccine-ready-use -next-month.

(216) "COVID-19-Impfstoff: Antworten auf häufig gestellte Fragen (FAQ)," Robert Koch Institut, last accessed August 27, 2020, https://www.rki.de/SharedDocs/FAQ/COVID-Impfen/COVID -19-Impfen.html.

(217) Michael Barry, "Single-Cycle Adenovirus Vectors in the Current Vaccine Landscape," *Expert Review of Vaccines* 17, no. 2 (2018): 163–73, https://doi.org/10.1080/14760584.2018 .1419067.

(218) Ewen Callaway, "The Race for Coronavirus Vaccines: A Graphical Guide," *Nature* 580 (April 2020): 576–77, https://doi .org/10.1038/d41586-020-01221-y.

(219) Barney S. Graham, "Rapid COVID-19 Vaccine Development," *Science* 368, no. 6494 (May 2020): 945–46, https://doi.org /10.1126/science.abb8923.

(220) Vincent A. Fulginiti et al., "Altered Reactivity to Measles Virus: Atypical Measles in Children Previously Immunized with Inactivated Measles Virus Vaccines," *JAMA* 202, no. 12 (December 1967): 1075–80, https://doi.org/10.1001/jama .1967.03130250057008.

(221) Hyun Wha Kim et al., "Respiratory Syncytial Virus Disease in Infants Despite Prior Administration of Antigenic Inactivated Vaccine," *American Journal of Epidemiology* 89, no. 4 (April

1969): 422–34, https://doi.org/10.1093/oxfordjournals.aje
.a120955.

(222) Cara C. Burns et al., "Multiple Independent Emergences of Type
2 Vaccine-Derived Polioviruses during a Large Outbreak in
Northern Nigeria," *Journal of Virology* 87, no. 9 (April 2013):
4907–22, https://doi.org/10.1128/JVI.02954-12.

(223) Z. Wang et al., "Detection of Integration of Plasmid DNA
into Host Genomic DNA Following Intramuscular Injection
and Electroporation," *Gene Therapy* 11 (April 2004): 711–21,
https://doi.org/10.1038/sj.gt.3302213.

(224) Barbara Langer et al., "Safety Assessment of Biolistic DNA
Vaccination," *Biolistic DNA Delivery* 940 (2013): 371–88,
https://doi.org/10.1007/978-1-62703-110-3_27.

(225) Norbert Pardi et al., "mRNA Vaccines—A New Era in
Vaccinology," *Nature Reviews Drug Discovery* 17, (2018):
261–79, https://doi.org/10.1038/nrd.2017.243.

(226) Peter Doshi, "The Elusive Definition of Pandemic Influenza,"
Bulletin of the World Health Organization 89, no. 7 (July 2011):
532–38, https://doi.org/10.2471/BLT.11.086173.

(227) Andreas Zumach, "Der verhängnisvolle Einfluss der
Pharmakonzerne," interview by Dieter Kassel, *Deutschlandfunk
Kultur* (Berlin), May 16, 2017, https://www.deutschland
funkkultur.de/weltgesundheitsorganisation-derverhaengnisvollee
influss.1008.de.html?dram:article_id=386282.

(228) Carsten Schroeder, "Schweinegrippe: Die Ruhe vor dem Sturm,"
Deutschlandfunk (Cologne), December 15, 2009, https://
www.deutschlandfunk.de/schweinegrippe-die-ruhe-vor-dem
-sturm.709.de.html?dram:article_id=88702.

(229) "Kanzlerin und Minister sollen speziellen Impfstoff erhalten,"
Spiegel, October 17, 2009, https://www.spiegel.de/wissenschaft
/medizin/schutz-vorschweinegrippe-kanzlerin-und-minister
-sollen-speziellenimpfstofferhalten-a-655764.html.

(230) Michael Fumento, "Why the WHO Faked a Pandemic," *Forbes*,
February 5, 2010, https://www.forbes.com/2010/02/05/world
-health-organization-swine-flu-pandemic-opinions-contributors
-michael-fumento.html.

(231) Clare Dyer, "UK Vaccine Damage Scheme Must Pay £120000 to
Boy Who Developed Narcolepsy after Swine Flu Vaccination,"
BMJ 350 (2015): h3205, https://doi.org/10.1136/bmj.h3205.

(232) S. Sohail Ahmed et al., "Narcolepsy, 2009 A(H1N1) Pandemic
Influenza, and Pandemic Influenza Vaccinations: What Is
Known and Unknown about the Neurological Disorder, the
Role for Autoimmunity, and Vaccine Adjuvants," *Journal of
Autoimmunity* 50 (May 2014): 1–11, https://doi.org/10.1016/j
.jaut.2014.01.033.

(233) Urs P. Gasche, "Corona: Medien verbreiten weiter unbeirrt statistischen Unsinn," *Infosperber*, April 26, 2020, https://www .infosperber.ch/Artikel/Medien/Corona-Medien-verbreiten -weiter-unbeirrt-statistischen-Unsinn.

(234) Michael Scheppe, "Risikoforscher erklärt: Das können wir gegen die Angst vor dem Coronavirus tun," *Handelsblatt* (Düsseldorf), March 10, 2020, https://www.handelsblatt.com/technik/medizin /gerdgigerenzer-im-interview-risikoforscher-erklaert-das -koennenwirgegen-die-angst-vor-dem-coronavirus-tun/25624 846.html.

(235) Christof Kuhbandner, "Von der fehlenden wissenschaftlichen Begründung der Corona-Maßnahmen," *Telepolis*, April 25, 2020, https://www.heise.de/tp/features/Von-der-fehlenden -wissenschaftlichen-Begruendung-der-Corona-Massnahmen -4709563.html.

(236) "12 Experts Questioning the Coronavirus Panic," *OffGuardian*, March 24, 2020, https://off-guardian.org/2020/03/24/12-experts -questioning-the-coronavirus-panic.

(237) "10 MORE Experts Criticising the Coronavirus Panic," *OffGuardian*, March 28, 2020, https://off-guardian.org/2020/03 /28/10-more-experts-criticising-the-coronavirus-panic.

(238) "Kritik an Corona-Berichterstattung der öffentlich-rechtlichen Medien," *Radio Dreyeckland*, March 27, 2020, https://rdl.de /beitrag/kritik-corona-berichterstattung-der-ffentlich-rechtlichen -medien.

(239) Nils Metzger, "Warum Sucharit Bhakdis Zahlen falsch sind," *ZDF Heute* (Mainz), March 23, 2020, https://www.zdf.de/nachrichten /panorama/coronavirus-faktencheck-bhakdi-100.html.

(240) Charlie Wood, "YouTube's CEO Suggested Content That 'Goes Against' WHO Guidance on the Coronavirus Will Get Banned," *Business Insider*, April 23, 2020, https://www.businessinsider .com/youtube-will-ban-anything-against-who-guidance-2020-4.

(241) John Oxford, "A View from the HVIVO / Open Orphan #ORPH Laboratory—Professor John Oxford," Novus Communications, March 31, 2020, https://novuscomms.com/2020/03/31/a-view -from-the-hvivo-open-orphan-orph-laboratory-professor-john -oxford.

(242) Rainer Radtke, "Anzahl der Todesfälle nach den häufigsten Todesursachen in Deutschland in den Jahren 2016 bis 2018," Statista, August 13, 2020, https://de.statista.com/statistik/daten /studie/158441/umfrage/anzahl-der-todesfaelle-nach-todesursachen.

(243) "Österreich: Regierungsexperten waren gegen Corona-Lockdown," *RT* (Germany), May 14, 2020, https://deutsch .rt.com/europa/102434-osterreich-expertenwaren-gegen -lockdown.

(244) "Dirk Müller: So schlimm wird es NOCH – und wer dahinter steckt! // Mission Money," YouTube video, 1:00:27, posted by "Mission Money," March 26, 2020, https://www.youtube.com/watch?v=Gf4yoHoEkCU.

(245) Stefan Homburg, "Finanz-Professor: „Das ist das größte Umverteilungsprogramm in Friedenszeiten"," interview, *Rundblick*, April 2, 2020, https://www.rundblick-niedersachsen.de/finanz-professor-das-ist-das-groesste-umverteilungsprogramm-in-friedenszeiten.

(246) Rubikons Weltredaktion, "Der Corona-Totalitarismus," *Rubikon*, March 30, 2020, https://www.rubikon.news/artikel/der-corona-totalitarismus.

About the Authors

Karina Reiss was born in Germany and studied biology at the University of Kiel where she received her PhD in 2001. She became assistant professor in 2006 and associate professor in 2008 at the University of Kiel. She has published over sixty articles in the fields of cell biology, biochemistry, inflammation, and infection, which have gained international recognition and received prestigious honors and awards.

Sucharit Bhakdi was born in Washington, DC, and educated at schools in Switzerland, Egypt, and Thailand. He studied medicine at the University of Bonn in Germany, where he received his MD in 1970. He was a post-doctoral researcher at the Max Planck Institute of Immunobiology and Epigenetics in Freiburg from 1972 to 1976, and at The Protein Laboratory in Copenhagen from 1976 to 1977. He joined the Institute of Medical Microbiology at Giessen University in 1977 and was appointed associate professor in 1982. He was named chair of Medical Microbiology at the University of Mainz in 1990, where he remained until his retirement in 2012. Dr. Bhakdi has published over three hundred articles in the fields of immunology, bacteriology, virology, and parasitology, for which he has received numerous awards and the Order of Merit of Rhineland-Palatinate. Sucharit Bhakdi and his wife, Karina Reiss, live with their three-year-old son, Jonathan Atsadjan, in a small village near the city of Kiel.